T0138339

The Experimental Self

synthesis

A series in the history of chemistry, broadly construed, edited by
Angela N. H. Creager, Ann Johnson, John E. Lesch, Lawrence M. Principe,
Alan Rocke, E. C. Spary, and Audra J. Wolfe, in partnership with the
Chemical Heritage Foundation

The Experimental Self

*Humphry Davy and the Making
of a Man of Science*

Jan Golinski

The University of Chicago Press | Chicago and London

Jan Golinski is professor of history and humanities at the University
of New Hampshire. He is the author of *Making Natural Knowledge* and
British Weather and the Climate of Enlightenment, both published by
the University of Chicago Press.

The University of Chicago Press, Chicago 60637
The University of Chicago Press, Ltd., London
© 2016 by The University of Chicago
All rights reserved. Published 2016.
Printed in the United States of America

25 24 23 22 21 20 19 18 17 16   1 2 3 4 5

ISBN-13: 978-0-226-35136-0 (cloth)
ISBN-13: 978-0-226-36884-9 (e-book)

DOI: 10.7208/chicago/9780226368849.001.0001

Library of Congress Cataloging-in-Publication Data
Names: Golinski, Jan, author.
Title: The experimental self : Humphry Davy and the making
of a man of science / Jan Golinski.
Other titles: Synthesis (University of Chicago. Press)
Description: Chicago ; London : The University of Chicago Press, 2016. |
Series: Synthesis | Includes bibliographical references and index.
Identifiers: LCCN 2015042392| ISBN 9780226351360 (cloth : alk. paper) |
ISBN 9780226368849 (e-book)
Subjects: LCSH: Davy, Humphry, Sir, 1778–1829. | Chemists—
England—Biography. | Scientists—England—Biography.
Classification: LCC QD22.D3 G65 2016 | DDC 540.92—dc23
LC record available at http://lccn.loc.gov/2015042392

♾ This paper meets the requirements of
ANSI/NISO Z39.48–1992 (Permanence of Paper).

Contents

Illustrations

Introduction

Ill-written lives are the best. Little vanities and weaknesses come out,
which better sense and better taste would shrink from detailing. Sir H. Davy's
life puts me in mind of many things, both bitter and sweet—occasions gone;
friends lost; time mis-spent; character misunderstood!
LETTER BY CHARLES BELL, 1839[1]

My imagination was vivid, yet my powers of analysis and application
were intense; by the union of those qualities I conceived the idea,
and executed the creation of a man.
MARY SHELLEY, *Frankenstein* (1818)[2]

How did someone become a scientist before there was such a thing?
How did one dedicate oneself to a profession that did not exist, or set
out to assume a social identity that was not yet available? At the beginning of the nineteenth century, there were no "scientists"—the word
had not yet been coined. No one was putting on a white coat or clock-
ing into the lab in the morning. Although many countries had long traditions of scientific learning, formal institutions devoted to it were few,
and they offered sparse opportunities for employment. Career paths
were not clearly laid out, so professional identities were not readily
available. Anyone who wanted to devote his or her life to the sciences
had to engage in a project of self-invention, one that required considerable creativity and resourcefulness.

This book is about Humphry Davy's experiments in selfhood, the
trials and tribulations by which he made himself into his own special

kind of scientific practitioner. Davy (1778–1829) came to be recognized as one of the foremost British men of science of the nineteenth century. Originally a country boy from a modest background, he was propelled through his remarkable accomplishments to a knighthood, a baronetcy, and the presidency of the Royal Society. In the first decade of the century he was a brilliant and celebrated lecturer at the Royal Institution in London. His chemical investigations led to the discoveries of sodium, potassium, and other elements and to the invention of the miners' safety lamp. He began his researches by elucidating the physiological effects of nitrous oxide and went on to study the powers of electricity to decompose matter. He wrote about geology, agriculture, and many other areas of applied chemistry. He was also a poet—a friend of Samuel Taylor Coleridge, William Wordsworth, and Robert Southey—and his own literary ambitions were apparent in his late-life works about salmon fishing and travel.

Davy was a self-made man—so much so that there was considerable puzzlement among his contemporaries as to what sort of person he actually was. He called himself a chemist, a philosopher, and a poet; but each of these terms had different connotations, and the combination of them in one individual was unique. One twentieth-century biographer dubbed him "the mercurial chemist," reflecting her sense that his personality was fluid and changeable, like the element mercury itself.[3] Contemporaries often designated him a "genius," which was another label he sometimes applied to himself. The idea of genius was much discussed in philosophical and aesthetic circles at the time, but the term does not suffice to capture Davy's identity. It simply raises the question of how a person came to be regarded as a genius and what it meant. Less flatteringly, Davy's contemporaries also characterized him as a "dandy" (someone who existed only for self-display) and an upstart (because of his vertiginous social ascent). Disdainfully viewing his rise up the ladder of rank, a Scottish aristocrat called him "a very little man . . . an absolute Quis" (a nobody).[4] The suggestion was that he was all pretense or performance, that there was no core to his being.

I think this charge fails to do justice to the complexity of Davy's self-formation, which was not merely a superficial adaptation. It is true that, in the course of his career, he had to shape his discourse, bodily deportment, and manners to be accepted in elite social circles. He developed rhetorical skills to make himself an outstandingly successful performer in the lecture theater. But there was more to him than a showman or a dandy. He was also engaged in a profound examination of his own subjectivity through both literary and experimental work. He self-consciously attached himself to ancient philosophical traditions that promoted self-discipline and bodily regimen. At the same time, he probed relations between mind and body in physiological investigations of respiration, galvanism, and animal electricity. In these inquiries, he treated himself as an object of scrutiny, sometimes taking considerable risks in the process. His willingness to submit to danger in his experimental investigations became a consistent part of his public reputation, complementing his self-presentation as a glamorous and passionate figure to the admiring audiences at the Royal Institution. In his writings, he also represented himself as emotionally susceptible to the beauties of nature and attuned to its sublime forces. He continued the process of literary experimentation and self-exposure in his idiosyncratic final work, *Consolations in Travel* (1830), in which he appeared in several guises.

One thing we can be clear about is that Davy was not a "professional scientist." Although the term is sometimes still used to describe him, it is completely anachronistic. Davy could not have aspired to become a "scientist" because the word was not used in his lifetime.[5] And he would have rejected the label of "professional" because a profession implied a status lower than his elevated social aspirations. In fact, he never received what we would consider a professional training in chemistry or any other discipline. We cannot solve the problems of Davy's identity by assimilating him to a narrative of scientific professionalization or disciplinary specialization.[6] He was forging his career at a time of important changes in science and society whose outcome could not be predicted. Nobody could have known that the long-term result would be the institutions we associate with modern science. We are familiar with a situation in which societies and journals are specialized by discipline, professional discourse is firmly demarcated from popularization, and the government provides substantial funding for research. Davy was not; and he could not have foreseen that this was where things were heading. To understand the shape of his life as he lived it, then, we have to avoid the teleological assumption that his aim was to conform to a social identity that only emerged in a later period. If we impose that kind of retrospective framework on his life, we ignore the creativity he needed to set his life's goals and overcome the obstacles in his way.

To illustrate the point, consider his place in this collective portrait, titled, "The Distinguished Men of Science of Great Britain Living in 1807–8"[7] [Figure 1]. Davy is close to the center, but not by any means the most prominent individual in this group. It would be hard to guess from the picture that, in 1807, he was at the height of his fame. That year he had just announced the dramatic discoveries of two new elements, and he was widely regarded as the greatest English man of science since Newton.[8] In the picture, his features are indistinct and he is a rather diminutive figure in the company of the men surrounding him. He is peering over the shoulder of John Dalton, who looks much more comfortable and secure. And this despite the fact that the scene is set in the library of the Royal Institution, the very establishment where Davy built his career and where his extraordinary popularity as a public lecturer contributed substantially to its reputation. The key point, however, is that the scene embodies a retrospective point of view. The collective portrait was published by the printer William Walker in 1862, more than half a century after the date at which it is set. Not only was there no occasion when all of these men were together in the library of the Royal Institution, but the individual portraits from which it was assembled were originally made at different dates. Davy's is taken from Sir Thomas Lawrence's painting of the early 1820s, which shows him just after his election as president of the Royal Society, with the safety lamp on the table beside him [Figure 2]. This was not how Davy looked in 1807–8.

Figure 1. "The Distinguished Men of Science of Great Britain Living in 1807–8." Engraving published by William Walker Jr., 1862. Set in the library of the Royal Institution, this scene was composed from individual portraits of the persons shown. The original drawing was made by J. Gilbert, J. L. Skill, and William Walker Sr. The engraving was published by William Walker Jr., along with his *Memoirs of the Distinguished Men of Science of Great Britain Living in 1807–8*, which included short biographies of the individuals portrayed. (See MacLeod, "Distinguished Men of Science.") Courtesy of the Chemical Heritage Foundation Collections. Photograph by Gregory Tobias.

Walker's elaborate picture, in other words, is a deliberate work of fiction. The modest place it accords to Davy is a sign of how his reputation had waned in the decades after his death. Walker was much more interested in the engineers who were Davy's contemporaries—men such as John Rennie, James Watt, and Matthew Boulton. He believed they deserved credit for what were coming to be seen as revolutionary advances in industry around the turn of the nineteenth century, so he made them the most prominent figures in the picture.[9] Chemists were given less prominence, although their science had enjoyed what some called its "Augustan Age" in Davy's time.[10] Walker's retrospective bias also affected his assignment of all these men to a single social category: "men of science." The engraving implicitly argues for the inclusion of engineers in this group. It was a point that could plausibly be asserted in the 1860s but was less likely to have been acknowledged five decades earlier, when the men in the picture would not have seen themselves as members of a single class or socialized together at all.

Davy's awkward placement in Walker's image shows how difficult his successors found it to fit him into the category of the man of science. And, given his fame and prominence in his lifetime, this difficulty implies that the category itself was less clearly defined then than it later came to be. At the beginning of the century, it was really not clear what it meant to be a man of science or who should be included under that heading. Living in that era, Davy was not able to assimilate to a well-established social identity. The project of fashioning himself as a man of science was one he had to undertake on his own initiative, with his own resources of creativity and imagination.

Walker's collective portrait reminds us just how much the scientific world was changing in this period. Davy lived at the time of what has sometimes been called "the second scientific revolution." It has even been said that the word "science" can first be applied in something like its modern sense to the activity that emerged around the turn of the nineteenth century.[11] Significant changes in scientific disciplines occurred at this time, and Davy played a central role in many of them. The revelation of startling new phenomena concerning gases, heat, electricity, and magnetism had profound impacts in many fields of re-

Figure 2. Comparison of detail from Figure 1 with the Thomas Lawrence portrait of Humphry Davy (Figure 5).

Working from Sir Thomas Lawrence's impressive "swagger" portrait of Davy, William Walker significantly diminished his prominence among his contemporaries. Davy appears here as a diminutive figure, surrounded by taller men, and partly obscured by the seated figure of John Dalton in front of him.

LEFT: Courtesy of the Chemical Heritage Foundation Collections. RIGHT: © The Royal Society.

search. The new sciences of physiology and biology began to investigate the nature of living things. Geology pulled back the veil from the deep history of the earth and life thereon. Chemistry was revolutionized in a series of radical theoretical innovations, beginning with the work of Antoine Lavoisier in the 1780s. These revolutionary changes were accompanied by new developments in scientific institutions, in Britain and other European countries. Science became the object of considerable public fascination, with middle-class men and women attending the performances of public lecturers and consuming new scientific publications. New institutions were founded, devoted to expanding scientific education or to sponsoring specialized research. New genres of books and periodicals were printed and marketed. Government patronage of the sciences was also extended, especially in areas with military applications.

It is important to keep in mind, however, that these institutional changes reached their culmination after the end of Davy's lifetime. In Britain, they really came to fruition only in the 1830s.[12] As we shall see, this had some unfortunate consequences for Davy's posthumous reputation. Before his death, a few of the leading advocates for reform of scientific institutions had come to see Davy as an obstacle to the changes they were promoting. And the perception was not entirely unfounded. The world in which Davy's life had unfolded was, in many respects, still that of the eighteenth-century Enlightenment. Personal patronage by members of the aristocracy was of crucial importance to Davy in building his career, while government support was almost entirely lacking. It is not surprising, then, that he remained captivated by a gentlemanly ideal of good character and modeled himself upon it. In recruiting the audience for his lectures in the early 1800s, he built upon the methods of Enlightenment public science, aided by his personal charisma and the display of spectacular new natural phenomena. As an author, Davy ventured into some of the new periodicals of the era, though he also published in the venerable *Philosophical Transactions* of the Royal Society, which had held to the same format for decades. During his presidency of the Royal Society in the 1820s, many of his colleagues thought Davy was keeping up the practices of his predecessor, Sir Joseph Banks, whose "learned empire" had controlled British science for forty years.[13] For this reason, some of them looked on him as a representative of the old regime, which had to be swept away in the interests of comprehensive reform.

When considering Davy's life, we have to try to balance what was new with what was old. Only in that way can we hope to recapture what it was like for him to live through a period of historical transformation. And, precisely because he lived at a time when significant changes were afoot, we need to be especially careful to avoid the lure of retrospective interpretations. Davy, like many of his contemporaries, was engaged in a project of making himself up. He was fashioning his identity while living his life, without any obvious model to follow and without knowing what shape the future would take.

In talking about Davy's experiments in selfhood, I want to emphasize three features of the way he forged his identity. In each case, my analysis draws on the work of recent scholars in the history of science and related fields. The first point is that Davy made himself the person he became. His identity was not determined by the institutional context in which he found himself. He used the resources provided by the culture around him, but he was not passively molded by hegemonic social forces. In this connection, I invoke the notion of "self-fashioning," which has been found useful by historians of early modern science.[14] In the sixteenth and seventeenth centuries, when modern scientific institutions were nonexistent, individuals active in the sciences had no ready-made social identity. They made their living in various occupations: teachers, writers, merchants, courtiers, doctors, members of religious orders, even mercenary soldiers. Some held positions in universities; a larger number were landed gentlemen of independent wealth. Even the very few who were employed by the ancestors of modern scientific institutions were not entirely shaped by their conditions of employment. They had to conduct themselves so as to attract patronage from aristocrats or ecclesiastics, to earn status in genteel social circles, or to gain standing in the worlds of publication or commerce. Personal connections and the behavior they demanded were more important than an individual's location in a learned society or a university. Identity was not conferred on individuals as a condition of employment. Rather, people had to fashion themselves to survive and flourish in the circumstances in which they found themselves.

To think of Davy in terms of self-fashioning, then, is to associate him with his early modern precursors, rather than with an ideal of the modern scientific professional. But it has been argued that even professional scientists in the early twenty-first century are far from institutionally molded stereotypes.[15] Contemporary scientists readily cross the boundaries of disciplines and institutions; they act entrepreneurially outside the formal expectations of their employment; they often pursue interests that are personal or sectional rather than communitarian. In these respects, they are departing from the universal ethos that has sometimes been supposed to govern their behavior.[16] They are forging their own identities, making themselves up with no less virtuosity and creativity than their early modern ancestors.

Davy, as we have noted, lived at a time of transition, when the early modern social world was giving way to something more like what we recognize as modern. He was, as one contemporary remarked, "the creator of his own fortune," but so were many others in his day and since.[17] There were also, however, factors in his situation that were specific to his own time. Among these factors were literary and philosophical writings that led many people to think about their own personhood and explore their own subjectivity more deeply than hitherto. This is the second theme I want to draw out by referring to Davy's experiments in selfhood. His project was more than a superficial matter of putting on an appearance or staging a performance. His public self-presentation resonated with introspective reflection, of which the evidence survives in his letters and notebooks. Here again, trends in recent scholarship have informed my interpretation of these more private aspects of Davy's personality.

Studies of the formation of the modern "self" have shown that this era witnessed an important change in individuals' sense of identity.[18] There was a deepening consciousness of personal subjectivity, associated with the artistic and philosophical currents of romanticism. German idealist philosophy, with which Davy was acquainted through his friend Coleridge, stressed the role of the individual subject in actively creating knowledge of the external world. The poets and artists of the romantic movement made a similar turn inward, building on the culture of sensibility that had taken root in the second half of the eighteenth century. They sought to develop a more authentic relationship to the natural world by cultivating their emotional and imaginative responses to it. Their conviction was that scenes of nature resonated with instinctual powers within themselves. Davy reflected these attitudes. A strong sense of interiority was evident in his inquiries into his own mental processes. He was heavily invested in exploring his own sensory and cognitive powers, his passions and imagination. At the same time, he cultivated a sense of the beauty and sublimity of the natural world, with which he claimed an intuitive sympathy. He displayed these aesthetic capacities as aspects of his genius, and he also understood them as integral to who he was.

Another aspect of the more profound sense of individual identity that arose in this period was a greater awareness of the fixity of such personal traits as gender, race, nationality, and class origins.[19] People of this time were increasingly coming to see these traits as inherent features of who they truly were. As historians of gender have pointed out, the differences between the sexes were being reconsidered.[20] They were increasingly thought of, not just as anatomical features but as pervasive determinants of personality and intellect. Men and women came to be seen as having entirely distinct intellectual, emotional, and physiological characters. At the same time, European writers distinguished more firmly than before among the various branches of humanity. The so-called races were believed to be demarcated by biological and psychological differences that went well beyond skin color.[21] Nationality and class origins were also increasingly viewed as fixed aspects of personal character, not to be shrugged off or altered at will. These trends also had a bearing on Davy's case. He was sometimes discomforted by people's tendency to regard personal qualities as fixtures of individual selfhood, especially when it came to gender and social class. His masculinity was called into question by his critics, in part because he had encouraged women to study the sciences. This criticism reflected the increased rigidity of gender roles and the insistence that men should be men in every respect. Davy was thought to have feminized himself by his association with intellectual women. In addition, his aspirations to reach a higher social class were often satirized. He was lampooned for aping the manners of his superiors, since it was assumed that his origins would inevitably stamp his character for life. Such personal qualities were not always to be kept private; they sometimes erupted into public discourse, satire, or gossip. This was another sign of the times, and an indication of how profoundly people's sense of who they were was changing.

Davy was a man of his time, in many respects. But his experiments in selfhood also reflected an awareness of ancient traditions of self-cultivation, which is the third theme I want to bring out. Recent scholarship has uncovered the importance and remarkable durability of these traditions.[22] The philosophical schools of late antiquity, especially Platonism, Epicureanism, and Stoicism, had all proposed methods for disciplining the body and mind. These practices have sometimes been called the *cultura animi* tradition, or (following Michel Foucault) "the care of the self." They had as their aim the attainment of a state of calm and contentment, the suppression of the passions, the sharpening of the mind, and the maintenance of bodily health. Recovered and put into practice by early modern intellectuals, they were thought to aid philosophical contemplation and study of the natural world. It was believed that the powers of reason and the acuity of the senses could be enhanced by training the body and the mind. To this end, philosophers and others adopted a variety of measures: regulating their diet and medications, exercising, solving mathematical problems, practicing sexual continence or celibacy, monitoring their bodily intake and output in daily journals, and so on.

In the seventeenth century, natural philosophers including Robert Boyle, John Locke, and Isaac Newton all practiced versions of these programs. They kept meticulous records of what they ate, the medicines they consumed, the mental and physical exercises they undertook, and other features of their regimen. Self-monitoring was deemed essential to mastering the bodily passions, which was a crucial condition for undertaking intellectual work. The circumstances surrounding an individual were also thought to be of importance because the quality of air and climate was understood to affect health substantially. Davy was aware of these practices, and he selectively appropriated some of them. Particularly as his health began to fail in his later years, he became preoccupied with his bodily well-being, taking meticulous care over what he ate, his exercise, and the climate to which he exposed himself. He had himself bled regularly and applied leeches to his head to alleviate what he believed was an excessively sanguine constitution.

He described this therapeutic program as one of "care & discipline & ascetic living," using terms that invoked ancient ideas of regimen.[23]

In other respects, however, Davy turned his back on classical notions of the care of the self. Far from rigorously suppressing the passions, he believed that they should be enlisted in certain kinds of scientific inquiry. His investigation of the effects of nitrous oxide saw him surrendering to the mind-altering qualities of the gas. He publicized his own susceptibility to the euphoria and bodily spasms it induced, even assuming the character of an enthusiast in the eyes of his contemporaries for this reason. This aspect of his experiments in selfhood saw him striking out in the opposite direction from the moderation and control that were central to the classical ideal. One could nonetheless say that he was engaged in a project of self-cultivation distantly descended from the ancient practices. Davy was still monitoring how he treated his body and recording how his mind reacted. He still had the aim of understanding the influence of bodily passions on the workings of the mind. In these ways, he was carrying on his own experimental project of self-formation. The ancient traditions of care of the self—though completely ignored in narratives of the emergence of the professional scientist—are highly pertinent to Davy's case.

This book is a study of the identities Davy assumed in the course of his life, not a comprehensive biography. There already are several biographies, the earliest ones written by those who had personal memories of him. The first was by John Ayrton Paris, a physician who used documents supplied by Davy's widow. His book was found utterly unsatisfactory by Davy's brother, John Davy, who believed Davy's reputation had been damaged by Paris's tendency to reproduce gossip and anecdotes about his eccentricities. John Davy's own—much more reverential—biography appeared five years after Paris's and three years before John brought out his brother's *Collected Works*. The issues of Humphry Davy's standing and identity have continued to resonate in subsequent studies, with each writer viewing his personality from his or her own vantage point.[24]

Rather than a traditional biography, I am offering a study of Davy's self-fashioning as it was viewed by himself and others in his milieu. A similar approach has yielded illuminating portraits of some of his contemporaries, including James Watt and Michael Faraday.[25] I have also been inspired by studies that explore the mental and physical lives of their subjects in tandem, a technique recently used to great effect in celebrated books about Charles Darwin, Alan Turing, and Stephen Hawking.[26] What all of these works have in common is that they call into question the traditional notion of the self that underlies conventional biography.[27] Classically, biography has been based on the idea of a unitary and consistent identity, which develops in some organic manner over the course of a person's lifetime. These studies, by contrast, portray a self that is a public construct, a persona adopted strategically for certain purposes, or alternatively a function distributed among the persons and things that surround a particular individual. They suggest that the boundary between the intimate and the circumstantial is permeable and that "high" intellectual activities may be closely connected with the "lowest" bodily functions—sex and digestion, for example. We are presented with a partitioned and fluid selfhood, a mosaic or network that is disassembled and reassembled periodically. I have tried to take a similar approach to Humphry Davy. I do not deny that his life had elements of continuity over its whole term, but I have chosen not to hang a biographical narrative on the assumption of a singular identity. Instead, I have explored the ways he developed distinct personae in particular times and circumstances. The overall assertion is that Davy was capable of assuming multiple identities through the course of his life and that this was one way in which he pursued his experiments in selfhood.

In adopting this approach, I have been intrigued to find that Davy himself sometimes reflected on the various identities someone might assume in the course of a lifetime. At several points in his notebooks, one comes across remarks that bear witness to this. Sometimes he sketched a fictional scenario or began to tell a story—for example, about a person emerging into self-consciousness or entering into a romantic relationship. June Fullmer noted the presence of these "moral tales" in Davy's early notebooks. She envisioned him as a young man sketching out these little melodramas of moral awakening, to which he gave such titles as "The Child of Education, or the Narrative of W. Morley," or "The Lover of Nature, or the Feelings of Eldon," or "The Dreams of a Solitary."[28] In each of the stories, an individual is subjected to trials, naive virtue is put to some kind of test, and a degree of maturity or self-understanding is achieved. At the same time, as Fullmer notes, Davy was playing with the psychological and physiological ideas he was absorbing in his studies, seeking to understand the processes of moral development in scientific terms.

On other occasions, Davy seems to have been meditating on the kind of narrative that could encompass the shape of his life as a whole. At one point, in a notebook dated to 1799–1800, he outlined what he called "The History of Passion—A Philosophical Narrative." The section headings he listed for the short piece give the stages of an individual's life, stages that in fact correspond to the adoption of distinct identities: "The infant, being of sensation. The youth, being of imagination. The lover. The social being. The logopatheist. The lover of money.—of science. The misanthropist. The lover of Nature. Recurrence of former feelings. The lover of future existence."[29] Davy was imagining that a person might pass through these distinct identities in the years between birth and death. The infant he saw as entirely subject to the senses, the youth driven by idealistic vision. After them come the discovery of love and the entry into society as the person matures. It is not clear what Davy meant by a "logopatheist," a word he seems to have invented; perhaps it denotes one who has achieved a balance between reason and the emotions. In any case, maturity also seems to bring with it the competing lures of money and science. This stage is followed by a withdrawal from society and reorientation toward nature, and life ends with a reversion to youthful emotions and a growing preoccupation with what lies after death.

Of course, writing this as a young man, Davy could not have known the form his life would actually take. But there is a striking correspondence between this sketch, at least in some of its features, and the way his career eventually unfolded. The correspondence is not exact, but it is fascinating that Davy himself envisioned the overall shape of his life in this manner. It confirms my decision to structure this book around a series of characters or personae he assumed at various times. So I begin with the passionate being of his youth, when Davy indulged his senses and imagination in his early scientific inquiries, especially into nitrous oxide. I label this persona "The Enthusiast." It is followed by "The Genius," concentrating on the period when Davy emerged as a prominent and charismatic figure on the stage of elite society as a lecturer in London. This public attention raised a moral quandary for him, since the allurements of fame and fashion were threatening to corrupt him. His identity as an emotional being was also under scrutiny at this point, as represented in the persona of "The Dandy." Commentators were pondering the degree to which he had become fixated with self-display, with the consequent moral decline and dissipation of his affections. His fame was nonetheless a condition of his success in the character of "The Discoverer." Davy leveraged his popularity with his audiences to bring off his greatest scientific accomplishments. But dedication to science still remained at risk of contamination by the competing ambition of wealth. Davy was able to raise himself above the latter imputation—at least after he had achieved financial security—by adopting the persona of "The Philosopher." As a philosopher, he disavowed all economic motives for applying science to practical purposes. Finally, he became "The Traveler," as he withdrew from English society to undertake a series of journeys on the Continent. He reawakened his early passion for scenes of nature, while also plunging into meditation on the prospects of continued existence after his death, which he feared was imminent.

These six personae or characters do not correspond to discrete episodes in Davy's life. They are overlapping rather than chronologically exclusive; Davy was capable of assuming more than one at a time. But they each emerged to prominence during a certain phase of his career. For this reason, I can give at least an outline of his life by describing them in order. The fact that the characters overlap significantly reminds us that there were strong elements of continuity running through Davy's life. Even though I choose to delineate these different characters, I do not wish to deny that he was in certain respects the same person from beginning to end. By reaching back and reinterpreting his past actions, and by reaching forward to anticipate the direction his life might take, he reinforced the dimension of continuity that offsets the discontinuities in his career. This process of looking backward and forward is an aspect of Davy's experiments in selfhood that should not be overlooked when reading the chapters that follow. Davy was highly adaptable to the circumstances he encountered. But he was also a singular human being, deserving of the respect accorded to personhood, which is premised on a continuity of existence from birth to death.

1. The Enthusiast

> [I] determined thenceforth to apply myself more particularly to those branches of natural philosophy which relate to physiology. Unless I had been animated by an almost supernatural enthusiasm, my application to this study would have been irksome, and almost intolerable.
>
> MARY SHELLEY, *Frankenstein*

Humphry Davy laid the foundations for his reputation as an experimenter and discoverer with his first substantial piece of scientific research. Just twenty years old, and working under the direction of Dr. Thomas Beddoes at the Medical Pneumatic Institution in Bristol, he devoted the year 1799 to an investigation of the properties and potential medical benefits of various gases. The most extraordinary effects were caused by the substance known as nitrous oxide, later dubbed "laughing gas." As the remarkable properties of this gas became apparent, Davy and Beddoes invited friends and family members to breathe it. They shared it with a series of distinguished visitors, including Gregory Watt and Thomas Wedgwood (scions of the great Midlands manufacturing families), and the poets Robert Southey and Samuel Taylor Coleridge. In his published report, Davy included testimonies by these people, who described in detail what had happened to them. They talked about experiencing a feeling of pleasure and euphoria, being overcome by spontaneous laughter and involuntary movements, and having a sense of enhanced vigor or physical strength. The descriptions were widely read and frequently commented on in the press; they made Davy's name as an experimental chemist. His brother, John, was right to say later that the nitrous oxide investigation had "established [Davy's] character as a chemical philosopher."[1]

What was that character, exactly? One of its aspects was an ability for painstaking research on chemical substances and their reactions. Davy showed himself a meticulous analytic chemist, capable of making precise quantitative measurements and reasoning accurately from them. He also devised a careful protocol for administering gases so that their physiological effects would be revealed. But public attention also focused on another feature of his character: his seemingly reckless use of himself as an experimental subject. The properties of nitrous oxide were established in large part by a lengthy process of self-experimentation. Davy breathed the gas on many occasions, sometimes in large quantities, over the course of several months. Alongside reports from others who had tried it, he recorded the gas's impact on his own senses and feelings. This showed a different aspect of his chemical persona: a willingness to use his body as an instrument of experimental investigation and to take the risks of pain or damage to his health that might follow. In these experiments, Davy displayed an inclination to make himself the subject of scientific inquiry and to endure the consequent discomfort and danger.

It was the nitrous oxide investigation that established these features of Davy's scientific persona. Although the episode has been extensively discussed by scholars, its contribution to the formation of Davy's identity has not been fully explored.[2] His work in Bristol has been treated as a late manifestation of the program of pneumatic medicine, in which reformist medical practitioners in several provincial towns used newly discovered gases to try to treat diseases. The controversy that erupted over this incident marks the end of the era when such experimental therapies carried the hopes of progressive intellectuals such as Beddoes and his friends.[3] My focus here will be on a slightly different issue: how the incident contributed to shaping Davy's identity as a man of science, or what his brother called his "scientific character."[4] Davy's use of nitrous oxide demonstrated its effects on his mind as well as his body. The gas inevitably disturbed the mental equanimity of the person breathing it. Thus, the published accounts displayed Davy's sensitivity to the influence of bodily passions on his mind. He appeared both as a fearless explorer of the physical symptoms produced by the gas and as peculiarly susceptible to its mental effects—as (to quote his brother once more) "of that temperament best adapted to be excited by it, and of a tone of mind best fitted to enjoy its excitements."[5] The image of Davy as a bold and risk-taking experimenter was thus complemented by a sense that he was also a man vulnerable to the sway of his passions, passions that he invested in his scientific work. The contemporary word for such a person—a word that was charged with multiple and strongly polarizing meanings—was "enthusiast." This complex and ambivalent character remained with Davy for the rest of his career. It was both an asset and a liability, a resource that he deployed in other episodes in his life but also a constant feature of public comment, both positive and negative.

None of these events were foreseen when Beddoes hired the young man from Cornwall in the autumn of 1798—the first of two life-changing opportunities offered to him before his twenty-third birthday. Beddoes was on the point of opening his Medical Pneumatic Institution in the Hotwells district of Bristol. The visionary and eccentric doctor had long nurtured ambitions of providing experimental therapies to the sick and making them freely available to the poor. For Davy, the offer of a job in Bristol promised to open up prospects he could only have dreamed of during his upbringing in Penzance, in the far southwest of England. At the age of fifteen, he had witnessed the death of his father, a farmer and wood-carver. He was educated at the grammar schools in his hometown and in Truro and then apprenticed to a local apothecary. Beddoes secured his release from the apprenticeship and raised his eyes beyond his original ambitions of becoming a provincial medical practitioner. In the two and a half years he worked for Beddoes, Davy came to think of himself—and to represent himself to others—as a chemist and an aspirant natural philosopher.

The transformation was a significant one. When he arrived in Bristol, Davy was almost entirely self-taught in science. He had read his way into chemistry from the textbooks of Antoine Lavoisier and William Nicholson and performed experiments with the aid of his sisters in his childhood home. He was not entirely isolated, however, having begun to make connections with other men who shared his scientific interests. Davies Giddy, ten years his senior, then deputy lieutenant of Cornwall and living just outside Penzance, was an early patron. Giddy had studied chemistry under Beddoes while the doctor was teaching at Oxford, and the two had formed a lasting friendship. Also important in bringing Davy to Beddoes's attention was Gregory Watt, a son of the chemist and industrialist James Watt, who was one of the most important financial supporters of the Medical Pneumatic Institution. Gregory had lodged with the Davy family during a visit to Cornwall in the winter of 1797–98 and formed a close friendship with the young man, based on shared interests in chemistry, mineralogy, and geology. At the same time, Watt's friend Thomas Wedgwood, son of the pottery manufacturer Josiah Wedgwood, was also visiting the town. Davy's acquaintance with these two men linked him to the circle of the Lunar Society of Birmingham, a group of provincial intellectuals whose devotion to the sciences and their practical applications had placed them at the forefront of the English Enlightenment.[6]

Beddoes was aware of these important personal connections. But he also recognized in his new employee a shared interest in speculative natural philosophy, especially as it concerned the fundamental processes of life. Davy had sent Beddoes an essay that the doctor would publish in 1799, in a locally printed collection entitled *Contributions to Physical and Medical Knowledge*. The essay, "On Heat, Light, and the Combinations of Light," showed that Davy was already developing an interest in the relations between "imponderable" entities and living matter. He proposed that light should be viewed as a chemical substance capable of entering into combination with others. What Lavoisier had named "oxygen," the fuel of respiration that sustained all living things, should therefore be renamed "phosoxygen," according to Davy, to acknowledge that light was one of its components. Once one understood that light played this essential role in respiration, Davy suggested, one could recognize it as the key that could unlock the chemical processes of life itself. Light, he claimed, must have an intimate connection with the basic functions of living things, and indeed with the higher mental processes. "On the existence of this principle in organic compounds," he concluded, "perception, thought, and happiness, appear to depend."[7]

The idea was attractive to Beddoes, who was easily drawn into theoretical speculation and was fascinated by the notion that life was basically a chemical process. Davy perhaps exaggerated a little when he wrote to his mother shortly after arriving in Bristol that Beddoes had "paid me the highest compliments on my discoveries and has in fact become a convert to my Theory."[8] But there was a genuine meeting of minds. For a while, Davy shared Beddoes's aim of using the Medical Pneumatic Institution to introduce enlightened reforms into medical practice. He too hoped to see therapies based on cutting-edge scientific research, with the benefits made available to all. But it did not take long for Davy's outlook and aspirations to diverge from his employer's. He was shortly disowning what he called his "infant chemical speculations" regarding light and ruefully admitting their lack of solid evidential foundations. As first nitrous oxide and then the phenomena of galvanism claimed his attention, he built a reputation as a resourceful and thorough experimenter, in contrast to Beddoes's more speculative and slapdash approach. He also ventured farther than the doctor was willing to go in arduous self-experimentation.

Beddoes wrote his own account of the nitrous oxide experiments, rushing into print before the end of 1799 with his *Notice of Some Observations Made at the Medical Pneumatic Institution*. Davy took his time, publishing his own *Researches Chemical and Philosophical: Chiefly concerning Nitrous Oxide* the following year. As Mike Jay has written, Beddoes's book was "full of vigorously ridden hobby-horses and generous in Shandean diversions."[9] The bulk of Davy's, by contrast, was devoted to chemical analysis of the oxides of nitrogen and studies of their effects on animals. In a volume of 580 pages, the personal narratives of breathing nitrous oxide, composed by Davy and his friends, took up fewer than one hundred. Leading up to them were quantitative studies of reactions, which allowed Davy to determine the composition of the nitrogen oxides, and a rather pitiless series of trials on animals. Doses of nitrous oxide, he reported, rapidly proved fatal to two kittens, one dog, two rabbits, two guinea pigs, and a mouse. On the other hand, a cat, another rabbit, another guinea pig, and another mouse survived shorter exposures. The implication was that "nitrous oxide acted on animals by producing some positive change in their blood, connected with new living action of the irritable and sensitive organs, and terminating in their death."[10] Such a conclusion highlighted the risks involved to the human subjects who took the gas, though Davy reassured his readers that humans could safely respire it for much longer than animals.

Toward the end of the book, Davy devoted about forty pages to his own experiences with nitrous oxide and other gases, followed by the shorter testimonies of eighteen other individuals who had tried it. Their reactions varied. For most, the experience was a pleasurable one. In fact, some claimed the gas gave them the most intense pleasure they had ever felt. But there were others who found it unpleasant or stressful. A common theme was the subjects' propensity to uncontrolled and energetic bodily motions, or at least the feeling of unaccustomed strength or vigor. Davy tried to control for the genuine effects of the gas by giving some of the subjects plain air or oxygen to breathe beforehand.[11] He also recorded their prejudices about it—whether they were predisposed to expect it to have an effect on them or not. Along with collecting the individual narratives, he published detailed instructions for preparing and breathing nitrous oxide. He described the various chemical reactions by which the gas could be produced and included a diagram of a gas-holder and breathing machine designed by his friend, the mineralogist and chemical manufacturer William Clayfield. Clearly the intention was to enable the results to be replicated elsewhere. By these measures, and by the various precautions and controls he introduced into the experiments, Davy was working to make a recognized contribution to scientific knowledge out of what were necessarily highly individualized experiences.

One reason why it was difficult to do so was that it was hard to describe the experience the experimenters were trying to reproduce. As Davy noted, "It is impossible to reason concerning [recollected sensations], except by means of terms which have been associated with them at the moment of their existence, and which are afterwards called up."[12] The obstacle, as he well understood, was that nitrous oxide yielded "sensations similar to no others, and they have consequently been indescribable."[13] Many of the other subjects in the trials also noted how hard it was to find language to describe what had happened to them. The problem particularly exercised the poets, Southey and Coleridge. Also lost for words was the future thesaurus compiler Peter Mark Roget, who as a young doctor took part in the Bristol experiments. Other participants reported that they had trouble even remembering what had happened, and they were severely challenged in trying to describe it. They recorded that their feelings were unparalleled in their prior experience and hence impossible to capture in ordinary language. One of Beddoes's patients, when asked how he felt, said, "I do not know how, but very queer." Another commented enigmatically, "I felt like the sound of a harp."[14] As one of Davy's friends, the young industrial chemist James Thomson, concluded rather hopelessly, "It is extremely difficult to convey to others by means of words, any idea of particular sensations, of which they have had no experience."[15]

Struggling to find a vocabulary in which they could share descriptions of their feelings, the subjects resorted to an eclectic mix of physiological and philosophical terminology. Davy set the tone here, recounting the first occasion on which he breathed nitrous oxide:

> The first inspirations occasioned a slight degree of giddiness. This was succeeded by an uncommon sense of fullness of the head, accompanied with loss of distinct sensation and voluntary power, a feeling analogous to that produced in the first stage of intoxication; but unattended by pleasurable sensation.

And, then again, on the following day:

> The first feelings were similar to those produced in the last experiment; but in less than half a minute, the respiration being continued, they diminished gradually, and were succeeded by a sensation analogous to gentle pressure on all the muscles, attended by an highly pleasurable thrilling, particularly in the chest and the extremities. The objects around me became dazzling and my hearing more acute. Towards the last inspirations, the thrilling increased, the sense of muscular power became greater, and at last an irresistible propensity to action was indulged in; I recollect but indistinctly what followed; I know that my motions were various and violent.[16]

What is notable here is an oscillation between third-person and first-person narration. Davy starts by describing his feelings as if they had occurred to someone else: "a slight degree of giddiness," "an highly pleasurable thrilling," and so on. But by the end of the passage, he is using the personal pronoun: "I recollect," "I know." Apparently, it was impossible to sustain the style he used at the outset, which distanced his feelings from himself. There was no way to evade the fact that he was immersed in the experience, so that it had to be described in first-person terms. It is interesting that Davy returns to the third person at the end of the passage, when he mentions the erratic and vigorous muscular actions produced by the gas. Here he implies that the bodily motions were not really his but imposed upon him irresistibly by an external agent. And indeed he records that he could not even recollect distinctly what he had done while under this influence.

In the language of Davy and the other subjects, talk of "sensations," "pleasure," and "thrilling" was paired with discussion of muscular motions and other anatomic changes. As several commentators have remarked, a particular physiological theory was under examination in the course of Davy's breathing experiments—namely, the "Brunonian" theory of the eccentric Edinburgh physician John Brown.[17] Beddoes was particularly partial to Brunonian ideas, having published an edition of Brown's works in 1795 and expressed the hope that they could provide a rationale for pneumatic therapy. Davy also referred to Brown's notion that human health and disease could be understood in terms of the contrary influences of stimulation and depression. The first question the two men asked of nitrous oxide was whether it was a stimulant (like oxygen) or a depressant (like carbon dioxide). But Davy soon became skeptical that the effects of the gas could be understood entirely in Brunonian terms. He found that it had a stimulating effect, like alcohol, but without the debility that would be expected to set in afterwards. There was no equivalent of a hangover.[18] In a letter of January 1799, before he began the nitrous oxide experiments, Davy claimed that Beddoes was already inclined to abandon Brown's theories. But the doctor's sympathies for Brunonian physiology remained strong enough to give rise to some degree of tension between the two men as Davy's work with nitrous oxide unfolded.[19]

This physiological or medical vocabulary was complemented by one drawn from aesthetics. Davy and his friends recounted what happened in terms of their sensations, noting whether they were pleasurable or painful. They also noted that pleasure and pain were often mingled, and they frequently used the word "sublime" to name these feelings. The sublime was a crucial term in eighteenth-century aesthetic theory and was repeatedly invoked in connection with nitrous oxide. Davy recorded that he experienced "sublime emotions" under the influence of the gas, and that the delight caused "has often been intense and sublime."[20] He suggested that the sublime occupied the point where extreme pleasure bordered on pain. And he made an analogy with physical variables such as heat, which could be situated on a comparable scale of sensation:

> Reasoning from common phænomena of sensation, particularly those relating to heat, it is probable that pleasurable feeling is uniformly connected with a moderate increase of nervous action; and that this increase when carried to certain limits, produces mixed emotion or sublime pleasure; and beyond those limits occasions absolute pain.[21]

Using the term "sublime" in this way, Davy was knowingly referring to a long and rich tradition of aesthetic analysis, as Sharon Ruston has recently noted.[22] The sublime was used to describe an aesthetic effect distinct from the appreciation of beauty, one in which a measure of awe or fear was mingled. Edmund Burke's *Philosophical Enquiry into the Origin of our Ideas of the Sublime and the Beautiful* (1757) defined the sublime as the effect produced by fearsome displays of the power of nature, violent phenomena (such as lightning and volcanoes), and great expanses of emptiness (such as the ocean or outer space). Davy connected the feelings produced by nitrous oxide with the effects of these natural phenomena in a footnote to his *Researches*, where he explained that sublime emotion, "with regard to natural objects, is generally produced by the connection of the pleasure of beauty with the passion of fear."[23] Nitrous oxide was said to lead to the same blend of pleasure and fear, produced (as Davy explained) by a heightened intensity of nervous activity. Such an explanation, referring both to aesthetic and physiological ideas, was consistent with eighteenth-century notions of the sublime. Burke had introduced physiological vocabulary into his own treatment of the subject, drawing on contemporary medical writings to explain the sublime in terms of nervous excitement and the contraction of muscular fibers.[24]

Since it combined the experiential and physiological dimensions, talk of the sublime was an appropriate way to describe feelings that had their source in the physical properties of a gas but also resembled responses to natural scenery or great works of art. Davy tried to determine how his experiences of nitrous oxide were altered by exposure to moonlight or dramatic natural scenery—the kind of phenomena that were more conventionally identified as sublime. Several of the subjects who breathed the gas at Bristol compared its effects to those of music. James Webbe Tobin, a poet and political radical whose brother was a playwright, likened his own response to his feelings about great literature: "It is giving but a faint idea of the feelings to say, that they resembled those produced by a representation of an heroic scene on the stage, or by reading a sublime passage in poetry when circumstances contribute to awaken the finest sympathies of the soul."[25] Lacking an agreed-upon vocabulary for objective description of their experience, the nitrous oxide subjects evoked aesthetic experiences in their past. The nearest they could come to recounting what had happened to them was to recollect their feelings when confronting an awe-inspiring landscape or witnessing a dramatic performance. But such descriptions admittedly stopped short of conveying the essence of the experience. A precise account of the physiological action of the gas remained elusive, and for this reason the claims of its effects were likely to be greeted with skepticism.

Nowadays, most people encounter nitrous oxide as an anesthetic in the dentist's surgery, though its use as a recreational drug has also been reported recently. Some scholars have argued that the Bristol group's breathing experiments pointed toward the use of other intoxicants, especially opium, by Coleridge and other literary men in the period.[26] Historians have also pondered the question of why nitrous oxide's anesthetic properties were virtually unnoticed in Davy's investigation and not further explored at the time.[27] Davy did record that the subjects sometimes failed to remember what had transpired during his experiments, and he speculated that this could make nitrous oxide useful in surgical operations, but the possibility was not systematically investigated. The gas was not recognized and established as an anesthetic until the 1840s. Part of the reason is probably that it was not inhaled in sufficiently high concentrations to produce anesthesia. But it was also the case that Davy and his friends were not predisposed to construe their experience in such terms. As Santiago Colás has put it, they were thinking more in terms of aesthetics than anesthetics.[28] The aesthetic vocabulary they inherited from their eighteenth-century predecessors allowed them to assimilate an unfamiliar experience to more familiar feelings about works of art or natural scenery. It also gave them ways to recount what happened in terms of pleasures and pains and to link these sensations to possible physiological causes.

The aesthetic vocabulary represented the experimental subjects as individuals overcome by their emotions; it therefore called into question their intellectual autonomy. Burke himself had used his notion of the sublime to challenge the idea—associated with the philosopher John Locke—that the mind could be isolated from the influence of the passions. In Burke's view, experiences of the sublime showed that states of mind could not be separated off from physical changes in nerves or muscle fibers. And yet, the Lockean ideal provided the philosophical foundation for the reporting of scientific observations and experiments. Such reports were meant to be composed by rational agents whose intellectual operations were independent of their passions. This supposition underlay the conventional protocols for experimental writing, which had been established more than a century before in such institutions as the Royal Society of London and had become generally accepted in the scientific world.[29] When reporting a series of experiments (with the air pump, for example), the prevailing style required a description of events as they were registered by the senses of a neutral observer. For the account to be as purely factual and objective as possible, the observer had to be autonomous, independent of any interference by preconceptions or passions. The ideal was a "modest witness," one who kept him- or herself out of the picture as much as possible. The observer was called on to practice a kind of self-suppression or self-abnegation. Hence, the stylistic conventions for experimental reportage developed in parallel with methods for purifying and clarifying the mind of the experimenter. These methods included mental exercises, self-medication, temperance, and sexual abstinence. The tradition of cultivating one's soul, or "care of the self," practiced by many early modern natural philosophers, sought to purge the mind of passions that could disturb the judgment or the working of the senses. Only if the mind were prepared by rigorous mental and physical discipline, it was believed, could it serve as a polished mirror that would perfectly reflect external reality.[30]

Any kind of self-experimentation departed from these expectations because it put the observer's intellectual autonomy at risk.[31] The root of the problem was that the reporters were also the instruments by which the phenomena were detected. Observers who used their own bodies to register physical forces found it hard to maintain their mental detachment. The conclusion was frequently drawn in the last decades of the eighteenth century in connection with such phenomena as electricity, galvanism, and "animal magnetism." To apply such forces to the body was to risk disturbing the workings of the mind and destroying the objectivity that was supposed to guarantee the reliability of the testimony. The problem was encountered by Davy's contemporaries who shared his interest in self-experimentation, such as the German natural philosophers Alexander von Humboldt and Johann Wilhelm Ritter, who used their bodies to measure the effects of electricity.[32] In their experiments, as in Davy's, the boundary between body and mind—which was also that between instrument and observer—was impossible to sustain. To report what had happened to them, self-experimenters broke with the traditional expectations of objective witnessing and surrendered their autonomy to the powerful forces acting upon them.

Davy, too, willingly embraced this procedure. He submitted himself entirely to the effects of nitrous oxide, aiming to experience those effects as fully as possible and then to describe them as best he could. He struggled to make his own description as accurate as the situation allowed, while guiding the other subjects to frame their narratives with similar care. But he recognized that unfamiliar feelings could never be entirely translated into public discourse. To give accounts of experiences that were not objectively observable, he had to abandon the traditionally detached persona of the scientific experimenter. Davy passionately committed himself to the role of self-experimenter. Far from emotionless, he trumpeted his emotional response to the gas. His susceptibility to its effects was precisely what qualified him—in his view—to bring his account before the world.

The nitrous oxide experiments were difficult to conduct and to report; they also raised problems of interpretation, especially in connection with the metaphysical issue of the relationship between mind and body. It was hard to decide what they implied for this long-debated and contentious question. Beddoes was inclined toward the materialist outlook, consistent with his radical political views.[33] For him, a chemical substance that led to peculiar sensations and emotions showed that mental phenomena could be caused by a physical alteration in the body. This implication provided support for the theory that identified the mind with the matter of the brain. Davy also was at first attracted by the materialist philosophy. Before he even met Beddoes, in the mid-1790s, he was speculating in a private notebook that mental powers might be due to a material substance, a kind of ethereal fluid, communicating between the brain and the muscles of the body. "Phaenomena which were formerly attributed [to] *psyche*," he proposed, "seem to be the effect of a peculiar action of fluids upon solids & solids upon fluids."[34] At that stage, he believed it could be proved that "the thinking powers depend on the organization of the body," a body that was nothing more than "a fine tuned machine."[35]

Perhaps unexpectedly, Davy's actual engagement with nitrous oxide led him away from materialism. The influence of Coleridge, who introduced him to the thought of Immanuel Kant in late 1799, may have confirmed his move in this direction.[36] Davy was soon vigorously crossing out the passages in his notebook that had lent support to the materialist view. Instead, he became attracted to the alternative philosophy of idealism, the notion that ideas in the mind rather than material things are the fundamental reality. One can see how the experience of breathing nitrous oxide might have suggested this notion. The gas tended to sunder the individual's consciousness from the surrounding material world, since it gave rise to sensations that had no external reference. The implications struck Davy with something of the force of revelation at the climax of an unprecedented binge on alcohol and nitrous oxide in November 1799. On that occasion, he recorded, he "lost all connection with external things." Then, with "the most intense belief and prophetic manner," he proclaimed that all that could be known was thoughts present in the mind. He cried out to a bystander, "*Nothing exists but thoughts!—the universe is composed of impressions, ideas, pleasures and pains!*"[37]

The idealist outlook, redolent of the philosophy of the early eighteenth-century Irish bishop George Berkeley, was further explored by Davy in a letter to James Webbe Tobin in March 1800. The surviving document is damaged, and a few crucial words are missing, but it is clear that Davy was musing on the question of what the idealist philosophy implied for the continuity of personal identity. "I have been puzzling myself to find out what people mean by external things," he wrote. He concluded that "when we say that an external world exists we mean nothing more than that ideas exist capable of modifying impressions." As for "<u>our identities</u> / for example self, our friends, all the people we know intimately, all the places we are well acquainted with &c," they are "connected with the possibility of our perceiving the impressions."[38] The point is expressed a little obscurely, but it seems that Davy was pondering an issue that had preoccupied some of the leading empiricist philosophers. If all that we really know is ideas currently in the mind, then how can we account for our belief that we and other people have a continuous identity over time? What allows us to remember experiences that are no longer present to the senses and to ascribe those experiences to ourselves or other individuals?[39] One can understand how breathing nitrous oxide, which had created unfamiliar and startling new ideas and had disrupted the normal working of the senses and memory, could bring these questions to the fore.

Indeed, the problem of personal identity seems to have been preoccupying Davy during the nitrous oxide experiments. In his notebooks from the time, one finds him moving between the laboratory record and spontaneous philosophical reflections. In the course of this movement, the continuity of personal identity often emerges as a theme. Puzzling over the problem, Davy composed a couple of fictional narratives about an individual who finds himself isolated in a wilderness and gradually becomes aware of himself and his surroundings.[40] At the end of the second narrative, he suggested that experiences of the sublime in nature could be intimations of personal immortality, reassurances that individual identity would continue after death.[41] In other notes, he conjectured that the individual sense of self might be formed initially by a child's experiences before birth. Perhaps "the conscious being[,] the indefinable something called I" had its origins in the fetus's perception of itself as distinct from the surrounding membranes and fluids in the womb?[42]

Davy was not the only person for whom the experience of nitrous oxide raised these issues. Others who breathed the gas also found their assumptions about identity and self-knowledge called into question. Particularly when they lost control of their bodily actions, they relinquished the self-presence normally expected of the knowing subject. Repeatedly, they recorded their inability to restrain themselves from laughing or making spontaneous movements. It is notable that the subjects' narratives (including Davy's) tended to switch to an exteriorized mode of description as they recounted making erratic theatrical gestures, jumping in the air, or running around the room. These actions were reported as if witnessed from outside because they manifested a loss of control over the body that implied a loss of sovereignty of the individual mind. As Lovell Edgeworth—who tried the gas under Davy's supervision—put it, nitrous oxide created the feeling of "not having any command of one'self [*sic.*]."[43]

The nitrous oxide experiments were unsettling for all of those involved. Breathing the gas led to novel and confusing feelings, which called for articulate description and at the same time thwarted the subjects' ability to speak authoritatively. Attempts to draw out the implications of the experiments yielded metaphysical conundrums about the relations between mind and body, and even existential anxieties about personal identity. In certain ways, the experience evaded attempts to impose upon it the framework of experimental protocol, reliable witnessing, and objective narrative reporting. In response, Davy abandoned the stance of a neutral witness and threw himself into self-experimentation. He turned out to be a highly susceptible experimental subject, and he fully recounted his susceptibility in his published account of the episode. He assumed, in other words, the persona of an enthusiast, one whose passions had been excited even at the expense of his judgment.

Davy's publication of the nitrous oxide experiments, and his way of presenting himself in connection with them, placed him in the line of fire of public controversy. As the accounts of the investigation circulated, they drew a good deal of attention, much of it skeptical and some downright hostile. Davy and his friends were criticized and mocked for their gaseous respirations, especially by conservative writers who identified Beddoes and his associates as political radicals. There were particular features of the Bristol group's work that such attacks could fasten upon, including the obvious parallels between nitrous oxide and other controversial healing techniques. The incident became something of a scandal, as the epistemological questions that Davy had confronted were overlaid with the social and political tensions of the day. This was a tumultuous period in British history, following the outbreak of the French Revolution and the repressive measures introduced against its supporters in Britain. Especially after the declaration of war against France in 1793, political animosity intensified between radicals and reformers, on the one hand, and those who supported the British government in its efforts to censor and suppress dissent, on the other. Joseph Priestley, the most famous English pneumatic chemist, had been driven from his home in Birmingham by a reactionary mob incensed by his support for the cause of reform; he emigrated to the United States in 1794. These circumstances shaped people's views about the work of the Medical Pneumatic Institution. They consolidated Beddoes's generally negative reputation among his contemporaries. They influenced opinion about nitrous oxide and its potential uses in medicine. And they contributed to forming Davy's image as a man of science in the public eye.

Two crucial terms captured the issues at stake in the controversy that swirled around the Bristol experiments: *imagination* and *enthusiasm*. Both terms related to the question of whether the purported effects of the gas were physiological in origin or, alternatively, due to some kind of collective hysteria in the Bristol group. Davy was acutely aware of the problem. At the beginning of his *Researches*, he had noted the complications that might arise from "the modifications of perceptions by the state of feeling."[44] It was understood that mental sensations could have their origins in an individual's imagination rather than in any external cause. Mesmerism, a topic of widespread public interest for nearly two decades by this date, was often held up as an example. When he first came to public notice, in Vienna and then in Paris, Franz Anton Mesmer had claimed discovery of a previously unknown physical force, "animal magnetism," which he could manipulate to maintain or restore health. By the mid-1780s, however, elite scientific opinion had reached the conclusion that the supposed cures were due to delusions among those who were susceptible to Mesmer's personal influence. The effects of mesmerism were ascribed to the patients' overactive imaginations. This did not prevent people from continuing to be fascinated with it as a possible treatment for various diseases. But it did redirect attention at the psychological rather than physical causes of its alleged therapeutic effects.[45]

Similar questions were raised about other popular modes of therapy. Patients had been treated with electrical machines since the 1740s by physicians, scientific lecturers, and devotees of pietistic or mystical religious movements.[46] In the hands of such fringe practitioners as James Graham, Gustavus Katterfelto, and William Belcher, electrical therapies were promoted to an extent many found implausible. Graham became famous in the early 1780s for offering infertile couples the use of the "celestial bed" in his Temple of Health in London, where they could supposedly benefit from electrical and magnetic influences conducive to sexual health.[47] Katterfelto, dubbed the "prince of puff," incorporated electrical therapy into his traveling show around the same time, alongside other scientific spectacles and magical tricks.[48] In the following decade, Belcher offered treatment with "intellectual electricity" at his premises in Oxford Street in London, although he never described clearly what this therapy was.[49] It seems plausible that the wilder manifestations of medical electricity relied a good deal for their effect on their patients' willingness to believe in them.

The 1790s also saw the emergence of the "metallic tractors," metal rods that were said to have curative properties when drawn across a person's body. The tractors were invented by Elisha Perkins of Plain-field, Connecticut. Like Beddoes and Davy, Perkins promoted his new therapy by circulating narratives of apparent cures, written in a factual and scientific style. Also, like Beddoes's pneumatic medicine, Perkins's tractors were said to offer patients a way to evade the vested interests of the medical profession and take their treatment into their own hands.[50] Interest in the tractors rapidly crossed the Atlantic, and by the end of the decade they were being subjected to skeptical investigation at the Bath General Hospital and the Bristol Infirmary. The prominent physicians William Falconer (at Bath) and John Haygarth (at Bristol) set up experimental trials in which patients were treated with fake wooden tractors, yielding results that were said to rival those produced by the genuine articles. Haygarth and his colleagues concluded that the curative agent was the patients' imagination, encouraged by the stagey performances that surrounded treatment and communicated from one person to another by a kind of epidemic hysteria.[51]

Beddoes and Davy were well aware of these experiments, in which their friend William Clayfield was participating, less than a mile from their own Medical Pneumatic Institution. In July 1800, Beddoes was said by Davy to have "always ridiculed the 'Tractors,' in common with all other reasonable men."[52] But Beddoes had earlier sent some "genuine" tractors for the experimenters at the infirmary to try, though they were not in fact used. Both men must have been aware of the implications of these experiments for their own investigation of nitrous oxide. The transatlantic tractors were judged to have no intrinsic potency but to owe their effects to the imagination. In view of this conclusion, how could one be sure that the effects of nitrous oxide were really due to its physical properties rather than to the fantasies of those who respired it? As Davy wrote in one of his notebooks at the time, it was inevitable that "much will be attributed to imagination" by those without direct experience of the gas.[53]

The suspicion haunted the investigation from the beginning. The novelist Maria Edgeworth, sister of Beddoes's wife, Anna, recorded her impression that individuals' reactions to the gas were determined

by their willingness to believe in it: "Faith, great faith, is I believe necessary to produce any effect upon the drinkers, and I have seen some of
the adventurous philosophers who sought in vain for satisfaction in a
bag of *Gaseous Oxyd*, and found nothing but a sick stomach and a giddy
head."[54] In his letter to Tobin of March 1800, Davy wrote jestingly that
"the Nitrous oxide & imagination are curing paralytic patients," as if
indifferent to which it was.[55] But it is hard to believe he could really
have been indifferent, especially while such phenomena as mesmerism
and metallic tractors were feeding contemporary anxiety about what
was called "enthusiasm." By this was meant indulgence of the imagination to the extent that it overpowered the restraining force of reason.
Conservative writers in the late 1790s viewed such intense passion as a
threat to the good order of society.

Both Beddoes and Davy used the word "enthusiasm" in connection with their work on nitrous oxide. In his published account, Davy
admitted that his "enthusiasm" after first imbibing the gas was perhaps strong enough to distort his recollection of its effects.[56] His notebooks show that his use of the word was not at all accidental. On the
contrary, he had made a note to himself to "Men.[tion] Enthusiasm"
at this point in his narrative. A few pages earlier, apparently writing
under the influence of the gas, he scribbled, "The imagination is influenced, ideal beings & realities mingled together." And then, a few
pages later, "I am a little of an enthusiast."[57] The word came naturally
to his pen when he wanted to acknowledge how his emotions might
have weakened his judgment. But the term was far from a neutral one;
it was highly charged in the political discourse of the time. The word
had been in common use since the mid-seventeenth century, when it
was applied to those who claimed to have received personal revela-
tions from God. Calling such people enthusiasts offered a way to cate-
gorize them medically, to classify them as mentally unbalanced rather
than genuine prophets. In the early eighteenth century, an enthusiast
was understood as someone who was antisocial, perhaps a captive of
mental delusions or religious frenzy, and refusing to comply with the
prevailing norms of politeness and civility. Some religious dissenters

adopted the term to describe themselves, feeling at odds with the expectation that they should mingle sociably with the population at large. By the end of the century, the term was being used by Edmund Burke and others for political as well as religious radicals. At this point, it suggested a kind of uncontrolled sociability, a shrugging off of inhibitions in small groups of individuals who surrendered to the same passionate extremism. Intellectual — as well as explicitly political — clubs and associations came under suspicion for cultivating enthusiasm. Such groups were thought to be allowing their passion for social change to overwhelm rational restraints, with potentially anarchic consequences for society at large.[58]

Davy was fully aware of these social connotations of the term he was using. In one of his notebooks from this time, he composed an essay, "On the Pretended Inspiration of the Quakers and Other Sectaries," in which he tackled the problem of distinguishing genuine religious inspiration from superstition or enthusiasm. The latter, he wrote, arises when people's spirits are agitated and pleasurable ideas succeed one another rapidly. Then an individual could imagine him- or herself favored by a divine revelation: "It is in the nature of these mental distempers ... that ideas are mistaken for realities."[59] As an example, Davy mentioned the case of Richard Brothers, a self-proclaimed prophet of the apocalypse, who had been confined to a lunatic asylum by the British authorities in 1795. This was exactly the kind of enthusiasm that so worried the government and its supporters, and the parallels with the symptoms of nitrous oxide inhalation were impossible to miss.

Beddoes and Davy thus realized that they were making themselves targets for attack. Beddoes's reputation as a political radical was already well established, and he had antagonized many in the medical profession by his strident calls for reform. Now Davy had brought to light a substance that stimulated the senses and the imagination, and he had shared it with like-minded progressive intellectuals. It was predictable that the Bristol group should be portrayed as a coterie of enthusiasts, whose indulgence in the mind-altering gas posed a danger to good morals and social order. Beddoes had even anticipated, when he published his own account of the episode, "that we might . . . earn from our contemporaries nothing better than the title of enthusiasts."[60] That was exactly what happened.

It did not take long for the nitrous oxide investigation to attract criticism and ridicule. The Manchester physician John Ferriar, who had earlier been a supporter of the Medical Pneumatic Institution, took Beddoes to task in the *Monthly Review* for failing to demonstrate his claims about the therapeutic value of gases.[61] An anonymous pamphlet called *The Sceptic*, published in Nottinghamshire in 1800, made fun of nitrous oxide, along with mesmerism and galvanism, as symptomatic "wonders" of an age of revolutionary upheaval. The Bristol "quacks-pneumatic" had befuddled themselves, the author suggested, in this all-obscuring fog. The reactionary *Anti-Jacobin Review*, which was closely allied to the government, launched a severe attack on Beddoes and his program. In May 1800, the journal printed a series of satirical verses, "The Pneumatic Revellers: An Eclogue," in which the Bristol group was ridiculed mercilessly and its gaseous therapy denounced as delusional. The poets Robert Southey and Anna Letitia Barbauld, who had imbibed nitrous oxide with Davy, were satirized in verses in the style of their own compositions. Beddoes came in for particular mockery. Building on the ideas of fellow radicals Erasmus Darwin and William Godwin, he was said to have suggested that breathing certain gases might allow human beings to postpone death indefinitely. This showed just "how far a philosopher may be carried by the force of a flaming imagination."[62] The *Anti-Jacobin Review* followed up with a damning review of Beddoes's *Notice of Some Observations*, in which the author was fingered as a materialist, whose "heated brain" was always chasing after the latest novelties instead of spending time in patient scientific investigation. His purported remedies were said to owe whatever effects they had to "the enthusiasm of the discoverers and the patients."[63]

Beddoes was the most direct target of these attacks, and his scientific reputation never really recovered in the years leading up to his death in 1808. Davy tried to distance himself from the doctor's methods and from the controversy surrounding him. In a letter to his Penzance patron, Dr. John Tonkin, in January 1801, he wrote with pride of his nitrous oxide experiments and concluded: "I cannot speak of the Pneumatic Institution and its successes without speaking of myself. Our patients are daily becoming numerous, and our Institution, in spite of the political odium attached to its founder, is respected, even in the trading city of Bristol."[64] But this was putting a brave face on it. The taint of scandal was not so easily shrugged off. In fact, it continued to color Davy's reputation even after he moved away from Bristol to a position in London in February 1801. The episode was still being brought up in criticisms of him two decades later, when he had reached the apex of his career. In 1824, the radical journal *The Chemist* reminded its readers that Davy (then president of the Royal Society) and Southey (then poet laureate) had both "fuddled" themselves with nitrous oxide.[65] Seven years later, Davy's first biographer, John Ayrton Paris, admitted that there was "something singularly ludicrous" about the whole incident.[66]

The politically motivated scorn with which the Bristol experiments were received did not kill off interest in nitrous oxide. But it confirmed a general impression that the effects of the gas owed something at least to the imaginations of those who breathed it. James and Gregory Watt and Matthew Robinson Boulton encountered serious problems when they tried to reproduce Davy's findings in Birmingham in late 1799. Writing to Joseph Black, professor of chemistry at Edinburgh University, James Watt admitted that preparation of the gas could be troublesome and it had "very different effects on different people."[67] Although Davy claimed in his letter to Tonkin that his experiments were being successfully replicated in Edinburgh, it appears that the findings there remained inconsistent. Joseph Priestley Jr. witnessed the results of the experiments in Bristol, but when he returned to his father's residence in Pennsylvania, he had trouble repeating them. James Woodhouse, professor of chemistry at the University of Pennsylvania, visited Davy in London in 1802 and took up experiments with nitrous oxide on his return to Philadelphia. He was inclined at first to ascribe the gas's effects to the power of imagination and changed his mind only when he tried again with his students a few years later.[68] Thomas Cooper, a Manchester chemical manufacturer who had emigrated to the United States with the older Joseph Priestley, gave his opinion about the gas in lectures at Carlisle College in Pennsylvania, around 1813. Cooper concluded that nitrous oxide "tastes sweet; produces dizziness or vertigo . . . [and an] inordinate propensity to laughter; increases muscular power; & braces the nerves." But he also admitted, "On some it has no effect, on some it has a pleasurable & on others a disagreeable effect."[69]

By the end of Davy's lifetime, the received wisdom was summarized in the fourth edition of Robert Macnish's *The Anatomy of Drunkenness* (1832): "Even the alleged properties of the gas have now fallen into some discredit. That it has produced remarkable effects cannot be denied, but there is much reason for thinking that, in many cases, these were in great measure brought about by the influence of imagination."[70] In his *Life of Davy* (1831), Paris concurred that "the gas in question possesses an intoxicating quality, to which the enthusiasm of persons submitting to its operation has imparted a character of extravagance wholly inconsistent with truth."[71] "Imagination" and "enthusiasm"—the two words that had emerged so prominently in the controversy following publication of the Bristol experiments in 1800— were still being used to characterize the nitrous oxide experience three decades later. It seems to have been impossible to detach the physiological properties of the gas from the circumstances in which they had first been investigated. And the suspicion that Davy and his Bristol colleagues had been biased by their overactive imaginations and collective enthusiasm could not be dispelled.

This characterization must have had some bearing on the fact that the anesthetic properties of nitrous oxide were not recognized until the 1840s. Before that, as one observer noted, the only practitioners who tried to use it for medical treatment were "a few ignorant or crafty empirics."[72] No experimental investigation of comparable seriousness to Davy's was mounted in the following three decades. Instead, the gas was deployed as a means of entertainment, either in public displays or occasionally in domestic settings. Private gatherings would sometimes be enlivened by intoxicating respirations. On the stages of London theaters, volunteers would be subjected to a dose and would amuse onlookers by their subsequent antics. At the Royal Institution, there were displays of the gas even before Davy arrived. Elizabeth Fox (Lady Holland) described a "ridiculous" episode in March 1800, when two members of the lecture audience were invited to share it. One of them was Sir John Coxe Hippisley, a prominent patron of the institution. The results in his case were said to be "so *animating* that the ladies tittered, held up their hands, and declared themselves satisfied."[73] This episode was perhaps one inspiration for James Gillray's famous caricature, "Scientific Researches!—New Discoveries in PNEUMATICKS!" published in May 1802 [Figure 3]. The print, however, features Davy himself as an assistant to the lecturer, so it may also reflect an incident in June 1801, after he had taken up his appointment at the institution. On the later occasion, he recorded, respiration of nitrous oxide "produced a great sensation."[74] Press coverage of the event, on the other hand, mentioned that at least one participant had become ludicrously intoxicated.[75] And Gillray's scatological portrayal also suggested that the results of inhaling the gas could be unpleasant and chaotic.

Such undesirable consequences may well have limited Davy's deployment of nitrous oxide in the theater of the Royal Institution, or at least curtailed his willingness to share it with his audience. But he did not entirely abandon it as part of his experimental repertoire. When the Scottish scientific lecturer James Dinwiddie returned to Britain in 1807, after several years in the Far East, he found that nitrous oxide was being shown at several places in London. He saw John Tatum demonstrate the gas at his lectures in Dalston Square. Tatum had no qualms about allowing his audience to participate, and Dinwiddie recorded that two gentlemen who did so "kicked and roared out."[76] Davy, on the other hand, whom Dinwiddie twice saw demonstrate the gas at the Royal Institution, kept it to himself. On both occasions, in January 1809 and March 1810, he breathed it from a bag in sight of the audience until it began to have some effect, but he declined to offer it to anyone

Figure 3. "Scientific Researches!—New Discoveries in PNEUMATICKS!—or—an Experimental Lecture on the Powers of the Air." Engraving by James Gillray, 1802.

The only contemporary image of Humphry Davy in the lecture theater of the Royal Institution is this famous satirical caricature. He is shown holding the bellows while assisting in the administration of nitrous oxide to a member of the audience, with catastrophic results. As June Fullmer first pointed out, many of the individuals in the picture can be identified (Fullmer, "Letter," *Scientific American*, August 1960). They include Sir John Coxe Hippisley (the unfortunate experimental subject), Count Rumford (standing, in profile, right), Isaac Disraeli, Henry Englefield, and William Sotheby. (For additional identifications, see http://www.rigb.org/our-history/iconic-images/new-discoveries-in-pneumaticks.) However, there is no single recorded incident that fits all the details of Gillray's depiction, and even the identity of the principal lecturer (probably Thomas Garnett but possibly Thomas Young) remains uncertain. It is likely that Gillray put together a composite scene to satirize the activities of the Royal Institution.

© National Portrait Gallery, London.

else.[77] Davy was unwilling to surrender nitrous oxide entirely; it remained an attribute of his public character as an experimenter. But he sought to avoid the disorder that could result from allowing his audience to breathe it, too. What he wanted to display was its effect *on himself*, to remind those who attended his lectures that he had established his name as a man of science by imbibing this substance and recording its effects.

What Dinwiddie witnessed was indicative of the way Davy assimilated nitrous oxide as part of his public scientific persona. The investigation had established his character as an experimenter in several respects that remained with him for the rest of his career. The meticulous and systematic nature of his inquiry allowed him to differentiate his own approach from Beddoes's, which he came to see as lacking in rigorous method. When he reflected on his Bristol experience in a letter of November 1801, he concluded, "Nothing can be done by the rapid generalizations of aspiring Genius. . . . But all things are effected by laborious, by patient investigation."[78] He thus found himself in agreement with the *Anti-Jacobin*, which had claimed that Beddoes was captivated by a flawed model of scientific genius that identified it with ambitious but ungrounded speculation. That was not the kind of genius Davy aspired to be.

On the other hand, Davy's own behavior with the gas had shown him as anything but a reasoning machine. The experiments had demonstrated that individual receptivity to nitrous oxide varied considerably, and he was obviously highly susceptible. This susceptibility enhanced the singularity of his genius, in his own view and in that of his contemporaries. On one of the first occasions he breathed the gas, in April 1799, he recorded a sudden conviction that "I was born to benefit the world by my great talents." A few pages later in his notebook he inscribed his name in large letters next to Newton's, as if it had now been revealed to him that his reputation would match that of England's greatest man of science.[79] At the same time, the path to such celebrity was shown to lie through arduous and risky self-experimentation. The Bristol bookseller and poet Joseph Cottle, who befriended Davy while he lived in the city, linked his recklessness as an experimenter with his enthusiasm. "No personal danger restrained him," Cottle declared, "he seemed to act, as if in case of sacrificing one life, he had two or three others in reserve on which he could fall back in case of necessity."[80]

Heedlessness of personal risk and a willingness to put his life on the line came to be characteristic of Davy's approach to scientific inquiry. Cottle noted that "he has been known sometimes to breathe a deadly gas, with his finger on the pulse, to determine how much could be borne, before a serious declension occurred in the vital action."[81] In the spring of 1799, Davy tried to breathe "hydrocarbonate" (hydrogen mixed with carbon monoxide). The trial caused him a temporary loss of consciousness and might easily have been fatal.[82] He seems to have regarded even suicide—a widespread preoccupation of the age since the publication of J. W. Goethe's novel *The Sorrows of Young Werther* in 1774—as a kind of experiment. In a notebook dating from his Bristol days, Davy wrote, "The application of the pistol to the head is a new exp[erimen]t concerning the results of which it is impossible to reason."[83] He seems to have imagined extending his strenuous trials of endurance to the point of self-destruction.

At the Royal Institution, Davy continued to cultivate an embodied mode of investigation that relied on self-experimentation and public self-display, and he manifested a reckless disregard for his safety while pursuing scientific inquiries. He used his bodily reactions to calibrate the strength of voltaic batteries. He risked injury in handling dangerous chemicals and incurred it at least twice in laboratory explosions. In 1807, he contracted a life-threatening bout of typhus fever while visiting the Newgate prison to advise on antiseptic procedures. He described his lengthy recovery and convalescence as "an experiment of 9 weeks."[84] Exploring the properties of a newly discovered compound in 1813, he created an explosion that injured his hand and left him temporarily blinded. Davy's reputation as a genius remained associated with a willingness to subject his body to strenuous trials in the course of his experimental inquiries. This aspect of his character had been established by his conduct of the nitrous oxide investigation.

In some respects, Davy pointed toward a style of experimental inquiry that would prevail later in the nineteenth century. Later decades witnessed many examples of self-sacrificing scientists, who would put themselves at great risk of injury or death in arduous expeditions or taxing laboratory investigations.[85] The ideal of objectivity that came to predominate during that period was identified with a willingness to endure bodily suffering, whether in the course of a polar expedition or in experiments on X-rays. But the parallel is not exact. The self-sacrifice idealized in subsequent decades was premised on a high degree of rationality and rigid self-control. Davy's readiness to put himself at risk in his experiments, on the other hand, was regarded as a feature of his passionate nature. Rather than being a deliberate strategy, it was seen as an aspect of his enthusiasm. As his brother, John, noted in his biography, Humphry Davy's intense intellectual exertions could almost be taken for a kind of derangement. His frenzy in the laboratory, when he was seized by a new discovery or a problem to be solved, was interpreted as the characteristic behavior of an enthusiast.[86] The chemist William Henry agreed, summing up Davy's character shortly after his death as "bold, ardent, and enthusiastic."[87] As the nitrous oxide incident had shown, Davy was capable of outpourings of passion that he was never able—and probably did not wish—to restrain. His persona was emphatically not that of an entirely rational being with his emotions under constant and rigorous control. On the contrary, his public image continued to be that of a zealous experimenter, one frequently swept along by his enthusiasm.

2. The Genius

Natural philosophy is the genius that has regulated my fate.
MARY SHELLEY, *Frankenstein*

Humphry Davy's appointment to the Royal Institution in London in February 1801 opened a new era in his life. The move allowed him to escape the political storms swirling around Thomas Beddoes and take up a more central role in the scientific world of the capital. He thereby freed himself from the taint of his Bristol employer's tempestuous reputation and established himself in the more secure and placid environment of a metropolitan institution. After arranging his appointment with Count Rumford—formerly Benjamin Thompson, a colonial loyalist from Massachusetts who had received his title from the elector of Bavaria—Davy wrote to his old friend Davies Giddy: "Thus I am quickly to be transferred to London, whilst my sphere of action is considerably enlarged, and as much power as I could reasonably expect, or even wish for at my time of life, secured to me without the obligation of labouring at a profession."[1]

Davy was right to expect that the job would bring him a considerable enlargement of his "sphere of action," though even he may not have anticipated quite how successful he would be in the metropolis. In the theater of the Royal Institution, the reputation he had built over the course of his work in Bristol was greatly enhanced. His extraordinary abilities as a lecturer earned him public recognition as a scientific genius. His courses filled the theater time after time, drawing an audience distinguished not only by its size but also by its elite social status and the large proportion of women it contained. Davy proved to be the most popular scientific lecturer in the building and the most famous in the city. He exploited the potential of dramatic chemical phenomena to satisfy the popular craving for visual spectacle. And he rode the wave of public fascination with science in general, and chemistry in particular, at this time. In fact, more than anyone else, he made the wave.

As he had foretold in his letter to Giddy, the appointment gave him as much power as a man of his age and background could reasonably hope for. He used the leverage of his popularity as a lecturer to raise his salary and renegotiate the conditions of his employment at the Royal Institution. He also used his charisma to advance his interests beyond its walls. As we shall see in a later chapter, he translated the authority he had won with his audiences into influence within the scientific community, inviting select groups to witness his laboratory experiments and staging demonstrations to resolve disputes with other specialists. All of this prestige was earned without his having to take up a "profession," as he originally intended to do as a medical practitioner. The Royal Institution promised Davy something better than professional status—namely, the autonomy that characterized a gentleman. It would have been hard for a provincial apothecary to achieve this standing by the career he had originally planned.

Eleven years' employment at the Royal Institution provided the facilities and resources to build a formidable reputation. But Davy's personal identity was not defined by where he worked. The institutional location offered him the venue in which he developed his charismatic persona, but the work of self-fashioning was undertaken on his own initiative and in his own manner. This may be a little surprising to modern sensibilities, just as Davy's evident relief in evading professional status is rather unexpected. We have become accustomed to the idea that institutionalization and professionalization went hand in hand in the development of modern science. In this respect, Davy's so-

Figure 4. Portrait of Humphry Davy by Henry Howard, 1803.

Painted shortly after his appointment as professor at the Royal Institution, this portrait by Henry Howard shows Humphry Davy with scientific apparatus and notebooks. It was exhibited by Howard at the Royal Academy in 1803 and purchased after Davy's death by his lifelong friend Thomas Poole. (See J. Davy, *Memoirs of the Life of Sir Humphry Davy*, 2:400.) Richard Walker (in *Regency Portraits*, 1:147) conjectures that the version now in the National Portrait Gallery may be a copy, in which case the location of the original is unknown. The NPG version is said to have been owned by John King, Davy's friend from his Bristol days.

© National Portrait Gallery, London.

cial identity was definitely not a modern one. He aspired to rise higher than professional status, and he did not define himself in terms of his institutional employment. Although he was required to deliver public lectures as a condition of his job, he proved much more creative and successful in this role than anyone had anticipated. And while some aspects of his research agenda were set by his employers, he carved out a large degree of independence in this domain too. On his appointment, he told Giddy, he had been promised "sole and uncontrolled use of the apparatus of the Institution, for private experiments."[2] It was a privilege he did not hesitate to exploit, even after he had resigned from his position.

Davy's social standing was the result of his personal charisma rather than his institutional situation. Although genius was often thought of as a purely intellectual quality, even a spiritual one, Davy manifested it as a feature of his physical embodiment, rooted in his individual speech, appearance, and behavior. In his lectures, he displayed the characteristics that allowed people to recognize him as a genius, and he tried to capitalize on this perception in his personal interactions—with sometimes uneven success. Many observers scrutinized his performances, trying to discern just what it was about his conduct or physical features that had brought him so much fame. Toward the end of his life, a group of younger men turned their skeptical eyes upon him. They were campaigning for root-and-branch reform of scientific institutions, and Davy attracted their animosity after he became president of the Royal Society in 1820. Among other things, the reformers wanted to raise the threshold of expertise for membership of the society and to open up opportunities for government employment of scientific specialists. They saw Davy as a tool of the aristocratic interest and an obstacle to their campaign. They were also suspicious of his dependence on the acclaim of public audiences and his apparent submission to female adulation and the currents of intellectual fashion. Their criticisms cast a shadow over Davy's final years, when many of the satisfactions he might have expected to follow from his achievements were denied to him, and the shine of his youthful genius had begun to tarnish.

As a young man, Davy had been preoccupied with the question of genius—its origins, its nature, and its capabilities.[3] His early notebooks are full of remarks on the subject. They convey the sense that genius is an inherent gift, a kind of emotional attunement to the sublime forces of nature, and utterly different from mere mechanical reasoning. In common with other thinkers on the subject during his lifetime, Davy sometimes talked of genius as a kind of spirit that came from outside oneself; rather than possessing it, one was possessed *by* it. In his last book he personified this genius as a spiritual apparition that communicates superhuman insights to the narrator. While still in his teens, he had composed a poem, "The Sons of Genius," one of the few poetic compositions he published during his lifetime. It appeared in the *Annual Anthology* in 1799. The eponymous sons of genius are said to be those who are inspired by reason and the delights of nature. They respond not merely to the beauties of the natural world but to its "great, sublime, and terrible" features:

> The sons of nature,—they alike delight
> In the rough precipice's broken steep;
> In the bleak terrors of the stormy night;
> And in the thunders of the threatening deep.

And their ambitions appear to be spiritual or otherworldly as much as philosophical or scientific:

> To scan the laws of nature, to explore
> The tranquil realm of mild Philosophy;
> Or on Newtonian wings sublime to soar
> Through the bright regions of the starry sky.[4]

In contrast, when the historian Harriet Martineau looked back on Davy's life and career in the 1840s, she described his genius as something imputed to him by his audiences. Davy, she wrote, "presented most strongly to the popular observation the attributes of genius."

Genius, on this account, was a performative quality, something enacted in displaying oneself to "popular observation." Trying to specify how Davy had acquired his reputation, Martineau listed his characteristics: "His ardour, his eloquence, his poetical faculty, the nature of his intense egotism, his countenance, his manners (before he was spoiled), and his pleasures all spoke the man of genius."[5] The themes were familiar from commentators in Davy's lifetime, whose words Martineau was echoing. His physiognomy and powers of speech were much discussed by his contemporaries. The genuineness of his manners and the strength of his ambition were weighed in the balance of moral evaluation. And those who witnessed his performances invariably noted his ardor and poetic sensibility. Watching him lecture at the Royal Institution in 1810, the French visitor Louis Simond remarked on how Davy's eloquence conveyed the intense feelings of someone who "has raised a corner of the thick veil, and untied one of the last knots of the great tissue of wonders." So masterfully did he embody these feelings that even "ordinary spectators experience an involuntary impulse of enthusiasm."[6]

The London lecture theater offered Davy a venue in which to exhibit his genius, communicating his enthusiasm in a way that compelled the attention of spectators without subjecting them to the risks of self-experimentation. This ability was apparent from his first lectures on chemistry in 1801. As Simond recorded, "though very young, and quite unknown at the beginning of the Institution, . . . his merit was soon estimated."[7] Despite his rustic origins, he revealed a talent for eloquence and a poetic turn of speech that engaged his listeners in even technical subject matter. And he never hesitated to exploit the potential of chemical phenomena for spectacular displays. According to a later commentator, "His exhibitions in the lecture room, were always sedulously contrived to interest and affect the multitude of the higher classes."[8] Nitrous oxide was one resource he used for this purpose. He also deployed a voltaic battery to create shocks, sparks, and loud noises. He applied electricity from the battery to ignite a little gunpowder or mercury fulminate and to draw figures on gold leaf by burning off the metal. After he discovered the alkali metals by using an electrical current to decompose their compounds, he featured sodium and potassium in his experimental displays. He threw lumps of them into water or onto blocks of ice, when they fizzed violently and exploded into flame. He also made a model volcano, "which threw out red hot lava at his call," to try to prove his theory that volcanic action was caused by alkali metals under the surface of the earth.[9] Tumultuous applause was said to have greeted this effect when it was shown in the theater.[10]

In the early months of 1802, Davy delivered his "Discourse Introductory to a Course of Lectures on Chemistry," which was soon hailed as an eloquent description of the potential uses of chemistry and the extraordinary powers of matter it had uncovered. In vivid terms, he discussed the properties of gases, the mysterious force of electricity, and the capacity of galvanism to reanimate dead matter. All of these powers, he told his audience, could be, "according to circumstances, instruments of comfort and enjoyment, or of terror and destruction."[11] A commentator noted that his use of the voltaic battery "involved the ladies in the prettiest terrors possible."[12] The combination of pleasure and terror was the familiar aesthetic effect of the sublime. By displaying his sensitivity to the sublime forces of nature and sharing them with his audience, Davy demonstrated his own genius. In his hands, chemistry yielded more resources to do so than the other sciences did. The Royal Institution employed several other lecturers, on astronomy, natural history, mechanics, and other subjects. But, as Simond noted, "These sciences are not . . . so fashionable as chemistry; they are not susceptible of any brilliant exhibitions; there is no noise, no fire, — and the amphitheatre never fills, but for Mr Davy."[13] The attractions of chemistry were obviously bound up with the attractions of the lecturer himself, the brilliant and glamorous young man from the provinces, with his poetic language and his feeling for the sublime. As Simond remarked:

> The elocution of this celebrated chemist is very different from the usual tone of men of science in England; his lectures are frequently figurative and poetical; and he is occasionally carried away by the

natural tendency of his subject, and of his genius, into the depths of
natural philosophy and of religion.[14]

These remarkable displays, in which the chemist was as compel-
ling as the chemistry, and a poetic sensitivity was brought to bear on
scientific subject matter, went well beyond what could have been envisioned when Davy was hired. At that point, the Royal Institution was
less than two years old and was still formulating its mission. Founded
at a meeting at Sir Joseph Banks's house in March 1799, it had taken
possession of its Albemarle Street premises three months later.[15] The
building was soon modified to serve the goals the new body had set
for itself, based on a scheme Rumford had been promoting for some
time. Its aims were quite different from those of the more venerable
Royal Society; Banks, as president of the latter body, made sure of that.
The Royal Institution was supposed to be dedicated not to advancing
the frontiers of knowledge but to teaching the application of the sciences to "the common Purposes of Life" and to diffusing knowledge
of "Useful Mechanical Inventions and Improvements."[16] Accordingly,
a lecture theater capable of holding a thousand people was constructed
as a central feature of the newly renovated edifice. A laboratory and
workshops were built in the basement, and a library and a museum
were also created to house collections of books and specimens. Dona-
tions from the wealthy and prominent people who committed themselves to support the foundation met the expenses of these facilities.
At the initial meeting, fifty-eight individuals subscribed for the substantial sum of fifty guineas each. They included the aristocratic landowners William Cavendish, fifth Duke of Devonshire, and George
John, second Earl Spencer, along with other members of the nobility,
and political leaders such as the antislavery parliamentarian William
Wilberforce. The subscribers were designated proprietors of the new
institution; they acquired lifelong rights to use its facilities and elected
a committee of managers to run it.

Davy's initial appointment, formalized on 16 February 1801, was as
assistant lecturer in Chemistry, director of the Chemical Laboratory,

and editor of the *Journal of the Royal Institution*. At first he was supposed to serve as assistant to Dr. Thomas Garnett, who had been recruited as a lecturer from Anderson's Institution in Glasgow. Davy was
granted a salary of one hundred guineas a year, furnished lodgings in
the building, and regular supplies of coal and candles. Understandably
excited, he wrote to his mother, Grace, as the negotiations were reaching their conclusion. The offer, he declared, was "of a very flattering
nature," coming as it did from "a very splendid Establishment," created
by "Count Rumford & others of the Aristocracy."[17] He emphasized the
elevated social connections of the institution and the highly honorable
nature of the appointment—"infinitely more honorable" than the position in Bristol, from which Beddoes had generously agreed to release
him. Interestingly, Davy recorded that agreement had been reached
for "my being in a short time sole Professor of Chemistry." Apparently
Rumford had already decided that Garnett's days at the institution
were to be numbered, and the doctor did indeed resign his post within
a few months. Davy also told his mother that his income would be "at
least five hundred a year." Although his initial salary was increased after
he took over from Garnett as lecturer, it still stood only at two hundred
guineas. The larger figure was presumably based on Davy's hopes for
what he could earn from private consultancy work, performing chemical analyses and other tasks for the proprietors.[18] It was clear, in other
words, that the formal conditions of his job would not stop him from
engaging in additional lucrative activities. Indeed, they provided the
social connections and public exposure that helped him do so.

Davy soon proved that the Royal Institution had made a wise investment in hiring him. Within a couple of years of his arrival, the body
faced a serious financial crisis. At the suggestion of its treasurer, Sir
Thomas Bernard, it adopted a series of measures to place itself on a
more secure footing. An annual subscription was started for those who
wished to use the library or attend lectures, making the most popular
facilities in the building yield regular revenue. Rumford had originally
planned a program of instruction in the applied sciences for working-
men, but this idea was abandoned, and the staircase originally de-

signed to allow the artisans to enter the lecture theater without encountering their social superiors was eliminated.[19] Instead, emphasis was placed on attracting the middle-class and elite audience that Davy had already begun to draw to his performances. He was asked to initiate a series on geology to complement his already popular lectures on chemistry. In addition to the large theater upstairs, a smaller one was constructed in the basement, and the wall dividing it from the laboratory was demolished. The aim was to allow about a hundred people to observe Davy working in the laboratory, and the space was later used to instruct medical students in applied chemistry.[20] The managers of the Royal Institution understood by this point that Davy was one of their strongest assets and that by giving him greater public exposure, they could bring in more funds. Without his contribution at this juncture, it is possible that the institution would not have survived.

Davy's unique value to the Royal Institution lay in his ability to draw the crowds. It was said that Albemarle Street was jammed with carriages on the days of his lectures, and the whole scene resembled a "noon-day opera-house."[21] His first series was said to have attracted "men of the first rank and talent, . . . blue stockings and women of fashion, the old and the young." Another observer recorded that the events were "attended not only by men of science but by numbers of people of rank and fashion."[22] The scientific lecturer James Dinwiddie, who first witnessed Davy's chemistry course in December 1808, described the audience as a "large and genteel company [of] six or seven hundred."[23] The presence of large numbers of women in the audience was frequently noted, though not universally approved. Davy was particularly encouraged by some influential aristocratic women among the subscribers to the institution. They included Lavinia, Countess Spencer; Georgiana Cavendish, Duchess of Devonshire; and Henrietta Ponsonby, Countess of Bessborough.[24] Some male critics worried publicly that science would be degraded and trivialized by subjecting it to the rule of female taste. But others were in favor of women taking an interest in chemistry, either because (like Davy himself) they approved of female education or on the grounds that "it keeps them out of harm's way."[25] When the American chemist Benjamin Silliman visited London in 1805–6, he wrote that scientific instruction of fashionable men and women was a good thing if it diverted them "away from scenes of amusement, where delicacy is perpetually violated."[26]

Davy's lectures offered a respectable alternative to such indelicate scenes of amusement. The literary scholar Thomas Dibdin wrote that the lectures "struck the learned with delight, and the unlearned with mingled rapture and astonishment."[27] Davy soon rose to the top of a large and competitive market for scientific lectures and other educational shows in London. There were many teachers offering public courses in chemistry and natural philosophy in the city, and chemical lectures were also provided in the major hospitals. The last two decades of the eighteenth century had seen a significant expansion of scientific activity, nurtured by public lecturers and other entrepreneurs, some of whom had built their careers in the provinces and were now discovering the opportunities offered by the capital.[28] The market was fed by the discovery of dramatic new phenomena in electricity, pneumatics, and galvanism, and by public fascination with such sights as manned balloon ascents. New collections of plants and animals were exhibited, and new technologies of display—including dioramas, panoramas, and planetariums—gratified the appetite for visual spectacles. These developments created the world of metropolitan public science in which Davy flourished in the first decade of the nineteenth century.[29]

When Dinwiddie returned in 1807 from a fifteen-year sojourn to China and India and set out to sample what public science had to offer in London, he attended lectures by John Tatum, George John Singer, Deane Franklin Walker, and others. He went to meetings of the Askesian Society (a group of chemists and other scientific men meeting at Plough Court in the City) and took in a lecture at the Spitalfields Mathematical Society (an artisanal organization in the East End).[30] He also visited various exhibitions and spectacles, including a Boulton-Watt steam engine, a purported perpetual motion machine, and—the new sensation—a "phantasmagoria" (a stage illusion in which a "ghost" was made to appear through the use of cleverly arranged lights and mirrors).[31] Dinwiddie's tour showed the extent and diversity of what existed in the way of scientific entertainment in the metropolis. Some of the offerings were inspired by Davy's success and intended to compete with him. The German immigrant chemist Friedrich Accum, who was for a short while Davy's assistant, resigned from the Royal Institution and within a few years was lecturing at the rival Surrey Institution. Singer began his own demonstrations of the voltaic battery at the Russell Institution in Bloomsbury in 1810. And the London Institution in the City provided facilities for Tatum's lectures. None of these competitor institutions lasted very long, and it is clear that Davy outshone all his rivals in terms of the size and social distinction of the audience he attracted. The competing efforts only paid him the compliment of trying to imitate him.[32]

Reviewing a biography of Davy published shortly after his death, a writer in the *Monthly Review* remarked that it was no denigration of his abilities to acknowledge how much he owed "to the address with which he kept himself almost constantly before the public."[33] The comment hits the mark. We do not diminish Davy's intellectual brilliance when we realize how much he depended on the setting in which he worked. He was a man of great talents, but he would not have been able to use those talents as he did had he not assumed a public role. His genius was essentially bound up with the social profile he attained. That profile shaped his work in the laboratory as well as in the lecture theater. The same reviewer recorded that, after the discoveries of the alkali metals, "the Laboratory of the [Royal] Institution was frequently crowded in a most inconvenient manner by persons of all sexes and ages."[34] Public witnessing of Davy's work in the laboratory certified his findings, and public acclaim spread the news of them. That acclaim in turn brought crucial support to the Royal Institution, the fortunes of which were otherwise precarious in its early years. According to Charles Babbage, it was by virtue of Davy's discoveries that the institution had risen "to a more prominent situation than it would otherwise have occupied in the science of England."[35]

Davy owed a great debt, then, to the Royal Institution, but it also owed a lot to him. It had allowed him an opportunity to address an audience drawn from the metropolitan elite, and by his own drive and talents he had captured their attention and earned their admiration. Their subscriptions bolstered the funds of the institution, and Davy was able to draw on these to help support his own research. He told those who heard his "Discourse Introductory" that "the man of true genius who studies science . . . will make use of all the instruments of investigation which are necessary for his purposes." Such a man, he went on, "is to a certain extent ruler of all the elements that surround him," capable of subduing to his will both common matter and "the ethereal principles of heat and light."[36] Casting himself in this role, he oversaw the construction of a one hundred-plate voltaic battery in 1803, which he displayed in the theater and later used to isolate sodium and potassium. In 1808, he launched a special subscription of the institution's supporters to fund an even larger instrument of two thousand plates. The spur was competition with French savants whose apparatus had been commissioned by the emperor Napoleon himself. And, when the battery was finished, it was hailed as a triumph of British public-spiritedness over French autocracy.[37] Dibdin wrote that Davy appeared in the theater "as the mighty magician of nature. . . . Begirt by his immense voltaic battery . . . [and calling] forth its powers with an air of authority, and in a tone of confident success."[38] Dinwiddie, who attended the lecture when the new instrument was unveiled, reported that Davy paid thanks to "the cheering voice of public freedom and popular applause."[39] The Scottish scientific journalist David Brewster, looking back from a vantage point a few years after Davy's death, agreed. Davy, he wrote, had created his own fortune: "Urged by the native impulses of a lofty ambition, he became the instructor of his fellow-citizens in the metropolis, and from their munificence and public spirit he received that assistance in his researches, which in other countries is proffered by the sovereign or by the government."[40] His genius, displayed in the metropolitan marketplace for scientific instruction, had been rewarded by an effusion of public support for continuing experimental research.

Brewster rather exaggerated Davy's independence in his posthumous encomium. Engaged at the time in writing a series of books on the heroes of science—including Galileo, Kepler, and above all Newton— Brewster was seduced by the image of the individual genius.[41] In his review of Davy's life, he declared that the young man had required "no titled patron" to cheer him on. This was far from the truth. In fact, Davy's titled patrons had played a significant role in his career by bringing him into the Royal Institution in the first place, recognizing his talents, and subscribing funds to support his work. Aristocratic and elite patronage was ubiquitous in British society in this period and re-mained fundamental to the workings even of the new establishment. For Brewster, and for Davy himself, this patronage was not incompat-ible with his having been "the creator of his own fortune."[42] The sup-port was freely given by private individuals, so it was thought to dig-nify both donor and recipient. This was what the British understood by "public freedom," which they contrasted with the degrading depen-dency cultivated by the absolutist French state.

As a young man, Davy was consistently successful at impressing powerful men and securing their support. He won the favor of Rum-ford and Banks, the moving forces behind the new institution, partly through the mediation of Thomas Charles Hope, an Edinburgh chem-istry professor who had witnessed his work in Bristol. Although Rum-ford departed for the Continent under a cloud in 1802, Davy continued to depend on the support of Banks, the most important patron of sci-ence in the period. It was Banks who proposed that Davy be pro-moted from lecturer to professor of Chemistry, with a further increase in salary, within a year of Garnett's departure. A few years later, ap-parently deferring to Banks's well-known opposition to upstart orga-nizations that threatened the hegemony of the Royal Society, Davy resigned from the newly formed Geological Society.[43] Another impor-tant patron was Sir Thomas Bernard—like Rumford, a colonial loyalist who had returned to England after the American Revolution. Davy's later work on agricultural chemistry attracted the support of major landowners such as the Duke of Bedford and Sir John Sinclair. He also had the backing of a group of proprietors who shared an interest in Roman Catholic emancipation. They included Sir John Coxe Hippisley, baronet and member of parliament, and Sir Henry Englefield, also a baronet.[44] His relations with these men extended beyond the confines of his institutional employment; they provided crucial connections as he built his career among the social elite of the capital.

As Davy rose in London society, he deployed his remarkable talents in settings outside the lecture theater, including gatherings at Banks's residence in Soho Square and at the country houses of prominent landowners. Within the Albemarle Street building, he opened the laboratory to spectators, including patrons of the institution, whom he invited to witness his work. The basement laboratory at the Royal Institution was a different kind of space from the large lecture theater upstairs, but Davy was said to have manifested the same qualities of genius in both places. In his biography of his brother, John Davy recorded that the laboratory was designed primarily for research: "there was no finery in it, or fitting up for display." At the same time, however, Humphry Davy was said to have maintained perfect openness in his investigations, never concealing any details from those who watched him at work. According to his brother, "he received his friends in the laboratory, and conversed with them on the objects of inquiry in progress; and however intensely engaged, he was always accessible."[45] Clearly there were differences between his activities in the laboratory and his performances in the theater, but it is striking that he welcomed spectators in both places. Laboratory work was also a kind of performance for him. And what he conveyed by the presentation was again his fervent self-investment in the process of scientific inquiry. John Davy used the words "zeal" and "enthusiasm" to describe his brother's demeanor in the laboratory.[46] Immersed in a kind of frenzy of experimentation, Davy displayed in the laboratory the same ardent genius he showed in the theater.

Whether lecturing, experimenting, or conversing with his patrons, Davy always seemed to be playing a part. His genius was made manifest by his poetic manner of speech, his physiognomy, and his mannerisms, even when he appeared to be immersed in introspection. Davy's poet-bookseller friend Joseph Cottle was one of the first to be struck by "the intellectual character of his face." Cottle explained: "His eye was piercing, and when not engaged in converse, was remarkably introverted, amounting to absence, as though his mind had been pursuing some severe trains of thought scarcely to be interrupted by external objects."[47] Subsequently, many other observers recorded their impressions of Davy's physical attributes, construing them as markers of his genius. As Davy exhibited himself, he in effect invited this kind of fetishistic appraisal. His body always seemed to be subjected to intense — even slightly obsessive — scrutiny. For this reason, when people speculated about his moral condition, they often did so by commenting on his body and its relationship to the physical environment.

This pattern of commentary was established when Davy moved from Bristol to London. Although he had rather pompously declared to his mother, "I will accept of no appointment except upon the sacred terms of *independence*," the fact was that by taking the Royal Institution job he was becoming entangled in the patronage networks of the government and the court.[48] Reformers and radicals, such as many of Davy's Bristol friends, had long been suspicious of such entanglements, a suspicion heightened by the government's repressive legislative and judicial actions in the 1790s. Sir Joseph Banks was personally at the center of an extensive network of establishment patronage, exploiting a close relationship with the monarchy and working through such institutions as the Royal Society, the Board of Agriculture, and the Royal Botanical Gardens at Kew. He had ostracized Beddoes when he discovered the doctor's sympathy for the French revolutionaries, and yet Davy was leaving Beddoes's employment to accept Banks's patronage.[49] Such a move was inevitably seen as questionable by the friends he was leaving behind and those who shared their political outlook.

The fact that Davy was taken up by London's social elite raised additional misgivings about his moral standing. At the time of his move to the capital, several of his friends expressed the worry that he would be led from the path of rectitude by the temptations of fame and the corrupting influence of fashion.[50] The poet Samuel Taylor Coleridge was one of them. In a letter to Robert Southey in October 1801, he expressed the hope that Davy's intellectual strength would protect his character from harm.[51] But, when he wrote to Samuel Purkis eighteen months later, he was still apprehensive of the danger. He suggested that two "serpents" surrounded Davy's "cradle of genius": "Dissipation with a perpetual increase of acquaintances, and the constant presence of Inferiors and Devotees, with that too great facility of attaining admiration, which degrades Ambition into Vanity."[52] Davy would need a Herculean strength of character to strangle both of these serpents, according to Coleridge. The question of whether he had in fact done so, or whether he had surrendered to popular applause, continued to be debated for the rest of his life, and indeed after his death. In 1811, Simond judged that Davy had not yet been corrupted by the adulation he had received. But, in later posthumous reviews of his life, it was often said that he had been. As an obituarist in the *Monthly Review* put it, "A fatal blight seemed to have fallen on the simple and genuine soul which he had brought with him from the country—and never afterwards . . . did he recover his natural character."[53] The applause of metropolitan audiences hailing Davy as a genius was thought by many to have compromised his moral purity.

When he first arrived in London, Davy himself pondered the ethical hazards of his new situation in letters to his friend John King, a Swiss exile and surgeon who was working with Beddoes at the Medical Pneumatic Institution. In June 1801, a few months after his arrival, he wrote, "The voice of fame is still murmuring in my ears—My mind has been excited by the unexpected plaudits of the multitude." He blamed himself for his "conceited egotism," which had led him to forget his obligations to his old friends.[54] Then in November, after a trip back to Cornwall had given him some distance from his London experiences, he wrote again to King, expressing his nostalgia for the ideas and affections of his Bristol days. In the capital, he wrote, he was "connected with new scenes & new beings. . . . Our existence is a round of varied feelings, in which nothing is permanent." He also admitted, "During the transition, irritability was induced & physical stimulation was recurred to. You will understand me & the explanation will plead as an excuse for me that I sometimes drowned moral sympathy in the vicious & vile physical sympathy."[55] The anxieties surrounding the move to London had led Davy to reflect critically on his own condition and apparently to confess to his friend some lapses in his sexual behavior.

While Davy associated moral corruption with sexual stimulation, other commentators linked the world of fashion and fame to the emasculating forces of luxury and decadence. In a series of articles in the *Edinburgh Review*, the Scottish Whig lawyer and journalist Henry Brougham reviewed Davy's Bakerian Lectures, in which he announced his major chemical discoveries to the Royal Society. In January 1808, Brougham remarked that it was a sign of Davy's great talents that they had "escaped unimpaired from the enervating influence of the Royal Institution; and indeed grown prodigiously in that thick medium of fashionable philosophy."[56] In July of the same year, he expressed relief that Davy had not named his newly discovered metals after the king, and so he had maintained his independence in "this courtly age." Davy's "political sentiments," Brougham judged, "are as free and as manly as if he had never inhaled the atmosphere of the Royal Institution."[57] When he came to write Davy's biography in his *Lives of Philosophers of the Time of George III* (1855), however, Brougham seemed to have changed his mind. There he decided that public fame had led Davy astray to "join its frothy, feeble current." The impressionable young man had been induced to try to merge science with fashion, producing a compound "somewhat resembling the nitrous gas on which he experimented earlier in life, having an intoxicating effect on the party tasting it, and a ludicrous one on all beholders."[58]

Brougham's choice of metaphors was an interesting one. Describing Davy's social environment as a "medium," "atmosphere," or "gas," Brougham was deploying the language of pneumatic chemistry. Joseph Priestley, one of the founders of the science, had linked the physical properties of gases to their moral qualities or virtues. "Good" air was thought to confer health and vitality, "bad" air disease and corruption. Similarly, the *Monthly Review* had spoken of "the noxious power of that climate of society in which such a mind as Davy's could have suffered immediate and incurable debility."[59] The suggestion was that the London atmosphere had been physically debilitating, that it had sapped the young Davy's manly strength. Such metaphors brought to mind a long tradition of theories about the effects of air and climate on people's character and behavior, theories to which Davy himself lent credence in his notebooks. They also evoked the Bristol pneumatic experiments and his surrender to the intoxication of nitrous oxide. To talk in this way about the hazards Davy faced in London was to evoke images of his body and the atmospheric conditions that impinged on it.

When he had mused about the attributes of the sons of genius in his youthful poem of that name, Davy had insisted that those who were distinguished in this way would rise above all mundane ambitions and temptations: "Above all earthly thoughts, all vulgar care; / Wealth, power, and grandeur, they alike despise . . ."[60] This poetic statement was to set very high—perhaps unrealistic—expectations for anyone who aspired to the role. For this reason, Davy's self-fashioning as a genius drew intense scrutiny as he made his way in the world. Precisely because the man of genius was meant to transcend earthly ambitions or material cares, all of Davy's career steps were subjected to searching moral evaluation, and he was often judged to have fallen short of the ideal. In addition, because he personified the figure of the genius in such a physically embodied way, moral appraisal often fastened on his body as its target. The man himself, in lecture theater and laboratory, drew attention to his body as a kind of instrument for detecting and channeling the powers of nature. He exhibited himself as a sensitive register of the sublime forces of the physical world, as he had done earlier when breathing potentially dangerous gases. It was not surprising, then, that his contemporaries pondered the effect of ethical hazards on that same sensitive body. Contemplating Davy as he plotted his course through London society, they wondered whether he had successfully navigated the buffeting winds that threatened him with moral harm.

The authority Davy acquired in the lecture theater of the Royal Institution was charismatic in nature. That is to say, it was individualized, embodied, and highly dependent on context. It was peculiar to him alone, communicated in the way he comported and displayed himself, and closely tied to a specific location, which made for difficulties when he tried to translate it to other kinds of relationship in other settings. Davy was extraordinarily successful as a lecturer to public audiences but less so as a teacher or mentor to individuals. As a writer, he explored several genres, as if struggling to establish his literary persona. The commanding presence he enjoyed in the lecture theater did not always come across on the printed page. And he had mixed success when he tried to assume a position of authority within the scientific community. Then, the scrutiny to which he had been subjected in the theater turned to censure and even ridicule. The physical attributes and gestures he had deployed in his performances were much less effective in nontheatrical settings. Even after he acquired wealth and titles of nobility, observers detected signs of his humble social origins in his speech, clothing, and manners. While Whig commentators worried that he had sold out to the establishment, Tories scorned him as an upstart. He was said to be inelegant in his dress, clumsy in his manners, and high-handed in dealing with those he thought of as his inferiors. It became a cliché to ascribe his moral weaknesses to deformations of character caused by early and excessive public acclaim. Thus, changes in his behavior were understood in terms of the long-standing narrative of his susceptibility to his emotions and the influence of his environment.

Davy never had students in any formal capacity, but he did act as a mentor to a few younger men. Some of these relationships resulted from or resembled familial ties. David Knight has gone so far as to write, "The childless Davy had four scientific sons."[61] It is true that John James Tobin, who accompanied him on his final travels, was his godson and the son of a lifelong friend, and Edmund Davy, who worked for a while at the Royal Institution and then made a successful career as a professor of chemistry in Ireland, was a cousin. The relationship with Michael Faraday, who rose from assistant to colleague to successor at the Royal Institution, was a more complicated one. In 1813, Davy hired him to assist in the laboratory and soon arranged to take him on a lengthy Continental journey. Faraday later recorded that he had been required on this trip to serve as a personal valet, a condition he found demeaning, though he placed most of the blame on the arrogant demands of Davy's new wife, Jane.[62] It must have been a source of disappointment to Davy that he never succeeded in imparting his own style of scientific persona to his quasi students. All of these men admired him, but none exactly followed in his footsteps. Although Faraday proved a successful lecturer in the very same theater, his style of performance reflected a more somber and less expressive personality. Davy's unique genius was not transmitted to the next generation.

Aside from Faraday, Davy's most important mentorship was of his brother, John, who was nearly twelve years younger and whose education he supervised with care. After finishing grammar school and then lodging with his brother in London for a while, John was sent to Edinburgh to complete his medical education in 1811. Humphry sent him money to help with expenses, cautioned him not to indulge in the "vice" typical of young men living away from home, and later recruited his aid in a dispute with the Edinburgh chemist John Murray.[63] Apparently worried that his brother was following a little too closely in his footsteps, Humphry wrote of his concern that John was experimenting on himself with digitalis and other toxins: "You may injure your constitution without gaining any important result[;] besides if I were in your place I should avoid being talked of for any thing extraordinary of this kind."[64] Humphry was obviously thinking of the effect of the nitrous oxide incident on his own reputation. He later wrote that an opening would be available for John to lecture at the Royal Institution if he wished to. But John's life path diverged from Humphry's. John pursued the medical career from which his brother had been diverted, published some respectable but not spectacular scientific works, and lived to an advanced age. He paid his brother the tribute of writing an adulatory biography and editing Humphry's *Collected Works*, but John never himself aspired to the standing of a genius.

Frustrated, perhaps, by his failure to nurture scientific progeny, Humphry Davy welcomed the opportunities for more extensive influence that opened up with his election as president of the Royal Society in 1820. But the experience was not a happy one. When he was obliged to resign from the office in 1827, after his first debilitating stroke, he reflected in a letter to his wife that he had derived satisfaction from the position and believed he had the support of most of the society's members, but "I had some enemies."[65] A year later, his recollections had soured even more, and he complained to Jane that he had been "used so ill by the public when I have labored most to serve them," citing his service to the society as an example.[66]

Davy's presidency was overshadowed by the prior tenure of Banks, whose autocratic rule over British science had endured for more than four decades. Banks had sponsored Davy's career in the Royal Society as well as in the Royal Institution. He supervised Davy's appointment as one of the society's secretaries in 1807 and presided over the award to him of its Copley Medal and subsequently its Rumford Medal. When Davy became president, some observers were convinced that he was trying to continue Banks's "learned empire," although poorly suited to do so. The Scottish geologist Leonard Horner wrote in April 1821 that he had visited the Royal Society because he was "curious to see how Davy would *look* the President." He reported that he had found

the Knight himself attired in a court dress, swaggering up to the chair. He had got on the coat & waistcoat of the court, but without either lace to his shirt or powder in his hair, & when he put on a very grotesque cocked Hat with a cut steel button he looked very like the porter at a shabby nobleman's gate towards the close of his livery suit. This imitation of Sir Joseph [Banks] is mighty ridiculous, & the more especially as Davy is the most awkward man [in] the world.[67]

Horner's comments struck a familiar chord in their preoccupation with Davy's physical appearance, manners, and clothing. The upstart Cornish chemist seemed to be aping the dress and deportment of his illustrious predecessor but lacking the gravitas to pull it off. There were substantive issues underlying the ridicule, given that Horner was allied with a reform-minded group within the society that resented what it saw as Davy's unwillingness to endorse the changes it was seeking. The group embraced members of the Geological Society, which had been abandoned by Davy in its struggles to gain independence from Banks's authority, and Cambridge mathematicians associated with the Analytical Society. Among their demands were government appointments for scientific specialists and higher standards of competence for election to the Royal Society.

David Philip Miller, who has written the most thorough account of Davy's presidency, sees him as caught between this faction and the aristocratic elite who dominated the organization, ultimately satisfying neither. Although he adopted some of the reformers' agenda at the beginning of his tenure, he failed to follow through with a comprehensive program. The changes he put in place, including the establishment of royal medals for meritorious work and the creation of the Zoological Society, were, according to Miller, "profoundly influenced by the mechanisms of aristocratic patronage which had made his own career possible."[68] As he extended favors to his allies within the society, Davy further antagonized the reform group. They were particularly incensed by his choice of his old friend John George Children over the Cambridge mathematician Charles Babbage for appointment to a vacant secretaryship in 1826.[69] Children had been a shooting companion of Davy's and his partner in a gunpowder-making business; Davy had promised to help him find employment when he faced financial hardship after failure of the bank owned by his father. After Davy pushed through the appointment, one of the reform party reported to Babbage, "The President has behaved infamously, full of his tricks and knavery of every description."[70] Babbage never forgave the insult. It contributed to what Miller calls Babbage's "extravagantly vituperative attitude," which he maintained even after Davy's death in barbed comments in his angry tract, *Reflections on the Decline of Science in England* (1830).[71]

Throughout this period of his career, Davy struggled to sustain what Miller calls "a great juggling act" that ultimately failed dramatically.[72] The assumption of responsibilities at this level required him to operate in a world where his reputation for genius did not convey authority, and where the techniques of the lecture theater were of little use. Davy had gained formal rank (a knighthood in 1812, a baronetcy in 1818) and—through his marriage—wealth; but his elite social status was acquired rather than inborn. He could never rid himself of the stigma of a lowly provincial origin. Those who studied his character often thought they could see through the superficial accoutrements of a man trying to appear better than he was. Sylvester Douglas, Baron Glenbervie, a Scottish politician of wide general learning and a fellow of the Royal Society, wrote of Davy that he was "a most eminent discoverer in chemistry. . . . But he will also be a fine gentleman[,] a man of rank and fashion . . . and to those who know him, and his disqualifications in respect of most of those ridiculous pretensions, he is a very little man." Ending with an Italian phrase that drew attention to Davy's dubious ancestry, Glenbervie, who was especially proud of his own genealogy, called him "oscuro figlio di non chiaro fonte" (an obscure son from unclear origins).[73]

One of Davy's actions that attracted these demeaning remarks was a proposal to include women in the audiences at evening meetings of the Royal Society. Banks used to hold informal gatherings for the fellows at his house at weekends. Davy continued the custom when he became president, and he floated the proposal that women should be invited to hear talks and participate in conversations on these occasions. He never envisioned their presence at formal proceedings of the society, but the suggestion was nonetheless flatly rejected by the fellows. On this matter, it seems, the young reformers saw eye-to-eye with the aristocratic old guard. Davy's suggestion likely reflected his experience at the Royal Institution, where women had been invited to attend lectures from the outset. But the notion that the Royal Society should follow the example of the younger establishment was thought by most of the fellows to threaten its reputation as a serious organization for scientific learning. They may also have been suspicious that Davy was looking for a way to enhance his authority within the more venerable society by capitalizing on his appeal to women. Given his knack of enthralling female audiences at the Royal Institution, he could well have been suspected of trying to reproduce the same effect in the circles of the Royal Society.

In any event, his failures to win support show that charismatic authority did not translate well to personal interactions outside the lecture theater. The eloquence and transports of emotion that had been assets at the Royal Institution did not help in the committees and hallway conversations of the Royal Society. Impressive equipment was not available in these settings to enhance his authority. And the youth that had contributed greatly to his reputation as a genius was receding in his fifth decade. Now, the talents that had drawn so much attention when he first appeared on the public stage seemed to be waning. Davy's personal attributes increasingly appeared as the superficial features of a social high-wire artist. In 1824, as his problems in the Royal Society began to mount, Davy was viciously attacked by the reactionary *John Bull Magazine and Literary Recorder*, in a profile in its series "The Humbugs of the Age." Davy, the magazine wrote, should stick to his crucibles and abandon the pretense of being a cultured gentleman: "The clothes of a gentleman do not sit easily upon him. . . . He smells of the shop completely." Readers of the article were reminded that Davy had started life as "a petty apothecary in some barbarous town in Cornwall. . . . He may be assured that he has still a gait and gesture, and habits and manners, nothing better than a village Ollapod."[74]

In the face of this kind of criticism, one can hardly blame Davy for recollecting nostalgically the adulation he had received as a younger man. He presented a kind of fantasy of this sort of charismatic authority in his final work, *Consolations in Travel: Or the Last Days of a Philosopher* (1830). In the series of dialogues between fictional interlocutors, he introduced a character called simply "the Unknown." The individual appears out of nowhere, encountered at the site of an ancient ruin. He lacks family connections or class origins, though it later turns out that he is English and an adherent of the established Anglican church. He has a natural air of authority, derived from his intellect, his voice, his ageless appearance, and his pseudoecclesiastical garb. He seems to demonstrate the possibility that one might command respect, independently of institutional structures or family background, by adopting a theatrical persona and making stage-managed appearances before captive audiences. Davy's fantasy of the Unknown looks like a projection of the genius he aspired to be.

When compared with the actual circumstances in which Davy made his career, such a figure can only be considered the product of wishful thinking. The Unknown claims to have attained independent wealth, allowing him to travel and mingle freely with the scientific elite. But Davy owed his reputation to the acclaim he earned in the lecture theater, while in paid employment of the Royal Institution. And he built his career through the patronage of wealthy aristocrats, who recognized his abilities. When he later attained gentlemanly autonomy and financial independence, he found that his reputation for genius yielded diminishing returns. Even while holding the rank of president of the Royal Society, he continued to be hampered by memories of his lowly origins and criticisms of his clumsy manners. Davy presumably thought of the Unknown as representing his latent persona, the true identity to which he acceded when wealth and title allowed him the freedom to attain a superior standing — after (in Miller's words) "the emergence of the butterfly that was Sir Humphry Davy, philosopher and well rounded gentleman of culture, from the chrysalis that had been Mr. Davy, ingenious chemist."[75] Although the attributes of gentility were thought by others to have eluded him, Davy still longed for the autonomy that he associated with genius.

The contradictions inherent in such a notion were, however, beginning to emerge, even before Davy's departure from the scene. Indeed, the aspirations represented by the Unknown soon came to seem rather preposterous for a man of science. The younger men who chafed under Davy's leadership of the Royal Society in the 1820s could recognize the tensions between his celebration of independence and his acceptance of aristocratic patronage. Subsequent men of science sought to develop their identities as employees of the state, as public servants rather than lackeys of the nobility. These men saw Davy's aspirations to gentility as having led him into demeaning compromises with aristocratic power. They portrayed him as captivated by an outmoded notion of genius and an obstacle to the rise of what was later called the professional scientist. The image was consolidated in the generation after his death and has continued to taint his reputation ever since.

3. The Dandy

I wished, as it were, to procrastinate all that related to my feelings of affection until the great object, which swallowed up every habit of my nature, should be completed.
MARY SHELLEY, *Frankenstein*

As Humphry Davy built his reputation, he took advantage of what people knew about his emotional susceptibility to the powers of nature and convinced them of his passion for scientific inquiry. By his poetic eloquence and dramatic displays, he captured his patrons' and audiences' support for chemical research. In effect, he achieved his youthful ambition of becoming a scientific genius through acting out the role in the lecture theater and the laboratory. The whole undertaking relied on his personal charisma, on his embodied presence in particular public locations. And there were inevitable risks associated with this mode of self-presentation. Someone who was so heavily invested in self-display might easily be judged superficial or vain. He could be regarded as morally defective or lacking in substance. We have already seen that Davy was subjected to this kind of critique. In this chapter, we consider what these criticisms reveal about changing understandings of gender in his milieu. We shall see that some critics thought his masculinity was compromised by his dependence on the admiring gaze of spectators, especially the many women who attended his lectures. He was satirized as a "dandy," a self-preening man whose identity was dependent on women's regard. It was a persona he never sought and surely resented, but it was symptomatic of the anxieties and uncertainties produced by shifting gender roles at the time.

Many observers recorded that Davy's performances at the Royal Institution appealed particularly to women. Louis Simond estimated that more than half the audience was female when he attended in 1810; he noticed that the women were much more likely to take notes than the men were.[1] James Dinwiddie mentioned that Davy flattered the women in his audience by complimenting them on their sensibility and discernment.[2] Davy's friend Robert Southey, writing under the pseudonym of "Don Manuel Alvarez Espriella," recorded that the men in the theater were mostly asleep or taking snuff, "while the ladies were all upon the watch; and some score of them had their tablets and pencils, busily noting down what they heard, as topics for the next conversation party."[3] Attempts were made by other lecturers in London to attract female attendees, but no one could quite replicate Davy's success, perhaps because no other lecturer could match the glamour of his personal appearance and poetic discourse. On the other hand, as we have seen, the Royal Society decided in the 1820s *not* to invite women even to weekend social events. The question of whether they should be included in scientific institutions continued to be debated even after Davy's death. In 1832, the founders of the new British Association for the Advancement of Science were obliged to consider the issue. The geologist William Buckland wrote to a colleague that women had to be excluded from formal proceedings to sustain an appropriately serious atmosphere: "[I]f the meeting is to be of scientific utility, ladies ought not to attend the reading of papers . . . as it would overturn the thing into a sort of Albemarle dilettanti meeting instead of a serious philosophical union of working men."[4] In his scornful reference to "a sort of Albemarle dilettanti meeting," Buckland likely had Davy's lectures at the Royal Institution in mind. They constituted a precedent that subsequent men of science were not sure they wanted to follow.

Davy deliberately chose to pitch his lectures at a largely female audience, a decision that reflected his conviction that women's education was desirable in an enlightened society and could appropriately include some knowledge of the sciences. This was a controversial position in the wake of fierce disputes about women's learning and their proper place in public life. Davy's identification with the cause of women's education, and the admiration he was accorded by female admirers, earned disapproval from conservative commentators and satirists. He was accused of pandering to uppity female intellectuals, including the notorious "bluestockings," whose nickname derived from the gatherings around Elizabeth Montagu in the mid-eighteenth century. Among Davy's contemporaries who earned this sobriquet was the young Scottish widow Jane Kerr Apreece, after the death of her first husband in 1807. When Davy became Mrs. Apreece's second husband in 1812, hostile gossip suggested that he had subordinated himself to an intellectually overbearing woman. The criticism hinted that his indulgence of women's intellectual ambitions was emasculating, that it had somehow unmanned him or confirmed an effeminate side of his character. It seemed to some observers that Davy's "dandyism" had been confirmed by his choice of a marriage partner.

These attacks reflected extensive and complex cultural shifts in Davy's lifetime. The gossip, scandal, and innuendo surrounding him were signs of much wider tensions between men and women in the period. As many historians have noted, the last decades of the eighteenth century witnessed the emergence of new ideas concerning the distinctive identities of men and women.[5] The attributes of gender came to be applied not just to the organs of reproduction but to all sorts of features of physiology, intellect, and character. Psychological traits and mental abilities came to be seen as bearing the imprint of an individual's masculinity or femininity. Many men thought women were less capable of the kind of higher reasoning practiced in the sciences and therefore should not be educated in scientific disciplines or admitted to its institutions. As arguments for excluding women, male ideologues deployed descriptions of the female mind that could be traced back to Aristotle and were now understood as having the ineluctable force of nature. They argued that women's grasp of the sciences would necessarily be constrained by their natural intellectual limits.

The increased prevalence of gendered notions of psychology did not mean that all individuals could easily be categorized in all aspects of their character. But it did imply that they were scrutinized more closely for what were taken to be masculine or feminine traits, and deviations from what was expected were more remarked and commented on. As the historian Kate Soper has written, "the interest of the period in sexual difference is also an interest in the instability of sexual identity, its dissembling and masquerade."[6] During the Napoleonic Wars, anxiety concerning effeminacy and homosexuality increased, as men who failed to fit the masculine stereotype were ridiculed, shamed, or prosecuted.[7] The drive to reinforce masculine ideals by ostracism or legal sanctions often emerges at times of military mobilization and threats to national security. Davy fell victim to some of this scrutiny. He was not accused of homosexuality, but his support for female scientific education led to suggestions that his masculinity had been compromised by submitting to brainy women or the gaze of admiring females. His manliness was called into question by those who did not share his commitment to women's learning, and the criticism followed him from his bachelorhood into the years of his marriage. In these circumstances, what might otherwise have remained private elements of his sense of self were brought to the surface and circulated in public discourse.

Delivering for the second year in succession a course of lectures at the Dublin Society in November 1811, Davy permitted himself some incidental remarks about women's scientific education. A knowledge of the sciences could refine women's powers of reasoning, he suggested, thereby curtailing "unhealthy sensibilities" and enhancing the pleasures of the mature years, after childbearing duties had been fulfilled. Female education therefore deserved the encouragement of the upper classes, rather than the unwise ridicule it had attracted. "The standard of the consideration and importance of females in society is, I believe, likewise the standard of civilisation," he concluded.[8] The sentiments were not uncommon in the enlightened circles among which Davy moved in the early part of his life. In Bristol, Anna Beddoes had drawn his admiration—and a strong measure of affection—for her intellect and poetic sensitivity. He dedicated one of his poems to her, in memory of the moonlit strolls they had taken together. Anna's father, Richard Lovell Edgeworth, and her sister, the novelist Maria Edgeworth, were both deeply interested in education, including that of girls. They advocated including the sciences in female education. After he visited the Edgeworth family home on a visit to Ireland in 1806, Davy exchanged correspondence with Maria on the subject.[9]

In proposing to open scientific knowledge to women, Davy could draw upon the precedents of many Enlightenment writers and lecturers. Authors such as Bernard de Fontenelle, Francesco Algarotti, and Jean-Antoine Nollet on the Continent and their British counterparts, including Benjamin Martin and James Ferguson, had addressed themselves to women in their presentations of natural philosophy.[10] Each had selected the content and style of their works with an eye to what they thought would appeal to the women in their audiences. Fontenelle's *Conversations on the Plurality of Worlds* (1686) had initiated a chatty, digressive style of scientific dialogue between male and female interlocutors. The mixed-sex dialogue proved remarkably popular and adaptable as a pedagogical genre of scientific texts, though the British tended to avoid the risqué element of flirtatious banter found in the French and Italian works. In his exposition of Newtonian optics, the celebrated *Newtonianism for the Ladies* (1737), Algarotti incorporated colorful imagery and occasional poetic interludes to render the subject more palatable. The English texts were generally more prosaic; but the form of a dialogue between a man and a woman was adopted by Martin to convey experimental physics and by Ferguson to teach astronomy. A conversational style, metaphoric figures of speech, and poetic digressions were among the means by which the pill of scientific knowledge was sweetened for female readers.

Davy followed his predecessors in certain respects in shaping a scientific pedagogy for mixed-sex consumption, but he emerged onto the scene at a time when women's education had become significantly more controversial. In the last decade of the eighteenth century, Catharine Macaulay's *Letters on Education* (1790) and Mary Wollstonecraft's *A Vindication of the Rights of Woman* (1792) were published, both of them upholding the cause of female learning. These works were answered by conservative tracts, such as Richard Polwhele's *The Unsex'd Females* (1798), which claimed that educating women threatened the normal relations between the sexes because it encouraged them to assume male roles and prerogatives.[11] It was along these lines that Davy himself became the target of conservative criticism for undermining women's subordination by teaching them science. The *Times* in 1813 alleged that he had been "making women and children troublesome, by the affectation and babble of knowledge."[12]

Aware that he was treading on dangerous ground, Davy seems to have adapted his presentation of chemistry to accommodate what he took to be women's interests. He conveyed the basic principles of the science as part of polite learning, but he did not encourage women to try to grasp practical details, given that they could not aspire to specialist or professional careers. One thing they should understand, however, was that scientific observations supported the belief in a wise and benevolent deity. Davy placed considerable stress on chemistry's connection with natural theology in his lectures, a link not often made by those who had previously taught that particular science. In his "Discourse Introductory to a Course of Lectures on Chemistry" (1802), he remarked on the chemist's perception, "in all the phenomena of the universe," of "the designs of a perfect intelligence." Along the same lines, he proposed that the science could be aesthetically appealing to women, especially by gratifying a sense of the sublime. Chemistry, he suggested, "must be always more or less connected with the love of the beautiful and the sublime; . . . [and] is eminently calculated to gratify and keep alive the more powerful passions and ambitions of the soul."[13] By stressing the spiritually uplifting aspects of chemistry and its compatibility with Christian doctrine, he tried to defuse the suspicion that he was encouraging women to nurture subversive ideas or ambitions to displace men from their occupations. His contemporary Leonard

Horner remarked, "Your chemists and metaphysicians in petticoats are altogether out of nature—that is, when they make a trade or distinction of such pursuits—but when they take a little general learning as an accomplishment they keep it in very tolerable order."[14] It was a sentiment with which Davy would likely have concurred.

The restrictions placed by conventional thinking on women's scientific learning were accepted by Davy's most renowned female disciple: Jane Marcet, author of *Conversations on Chemistry* (1806). Marcet was the daughter and the wife of Protestant immigrants from Geneva. She and her husband, Alexander, a prominent London physician who was himself teaching chemistry at Guy's Hospital, were good friends of Davy. *Conversations* was published anonymously, but Marcet disclosed her sex in the introduction. She explained that the project was inspired by Davy's lectures, and the plan was to provide women who attended them with the further information they would need to understand them more fully. She admitted that she had no specialist knowledge of chemistry but insisted nonetheless that the subject was a proper one for women to learn as part of polite culture. The book adopted the dialogic form pioneered by Fontenelle, unfolding through a series of conversations, in which an instructor, "Mrs. B.," led her girl students, "Emily" and "Caroline," through the basics of the science. The author explained that the conversational style was an appropriate mode of instruction for the female mind, unaccustomed as it was to dealing with abstract ideas.[15]

Marcet took considerable care to situate herself in a position that would appear respectable for a female author purporting to teach chemistry to female readers. She acknowledged that it was inappropriate for a woman to enter "into the minute details of practical chemistry." Her fictional pupils were enjoined by their teacher to refrain from pedantically using chemical terms in everyday conversation, which could only attract ridicule, and not to think of studying pharmacy, which "properly belongs to professional men."[16] The students were introduced to the world of science within a clearly delineated domestic setting, in which only small-scale experimentation was possible. They were shown the scene of a laboratory and invited to admire its powerful instruments but told not to expect to handle them. The laboratory was to remain an exclusively male domain. They were also led to reflect upon the sublime spectacle of nature, itself a kind of laboratory in which power was wielded not by the man of science but by God. Marcet encouraged her female readers by assuring them that "a woman may obtain such a knowledge of [chemistry], as will not only throw an interest on the common occurrences of life, but will enlarge the sphere of her ideas, and render the contemplation of nature a source of delightful instruction."[17]

Marcet's *Conversations on Chemistry* was enormously successful and evidently embraced by Davy as a worthy complement to his lectures. In a letter to her husband in May 1806, Davy dubbed Marcet "our tutelary Lady in Chemistry."[18] When he went to Dublin to deliver his first course of lectures there, in 1810, he reported that the book was popular and was being recommended by the prominent Irish chemist Richard Kirwan to all the ladies of his acquaintance.[19] Marcet's work was famously said by Michael Faraday to have inspired his interest in chemistry when he read it while working as a bookbinder's apprentice. There were many subsequent editions and foreign translations, and the book was pirated dozens of times in the United States. Insofar as it reflected the approach of Davy's lectures, it showed the limits set by contemporary prejudices about female participation in the sciences. Women were to be allowed a general view of the basic principles of chemistry, but it was not expected that they would engage in its various fields of practical application. They were encouraged to appreciate the signs of God's design revealed by chemical investigation but not to probe (for example) the chemical basis of living things. Along the same lines, Davy introduced eloquent digressions on natural theology into his lectures, to the great applause of his audiences. Simond even suggested that he should make more of his poetic inclinations in this respect, "abandon[ing] himself more naturally to his spontaneous feelings."[20]

Not all of Davy's female auditors were impressed by this technique. The poet Eleanor Anne Porden, later to marry the polar explorer Sir John Franklin, was a teenage girl when she first attended Davy's lectures. Largely taught by her father and having been given the run of his library, she rejected the notion that any branch of the sciences should be forbidden to women. In her account of Davy's last lecture at the Royal Institution, in April 1812, she expressed appreciation for the fact that he had toned down the mannered oratory displayed on previous occasions. The presentation, she wrote, was "delivered in a modest and proper and natural manner without any of the pomposity and rolling about which he has affected lately."[21] Porden took note of Davy's argument that scientific achievements were the sign of a strong and civilized nation, but she was relieved that the topic was discussed in a relatively cool and unemotional style.

Nonetheless, the suspicion that Davy was pandering to women was held by many of the men who observed him. He himself worried about the implications for his public image. In an undated remark in one of his notebooks, he asked, "How many fools who cannot produce an axiom are taken for men of genius because they are followed by old women[,] by young women[,] by girls & by young boys[?]"[22] The accusation that he was serving the tastes of female onlookers did not substantially damage his image as a genius, but it earned him a reputation for vanity and subservience to women's fashion. It also compounded the sense of a feminine streak in his character, a suggestion that had arisen earlier in connection with nitrous oxide. Davy had recorded that he was hesitant to allow women to breathe the gas because he was worried that its effects would overwhelm them. When he did try it on several female subjects, he confirmed that women were generally more susceptible than men. On the other hand, his own susceptibility to nitrous oxide was evident in his whole account of the investigation. He was his own best subject precisely because he was so vulnerable to the gas's effects. If responses to the gas were to be differentiated on grounds of gender, then, it appeared that Davy had a pronounced female side to his personality. His emotionalism and flights of poetic fancy in the lecture theater seemed to confirm it.

Suggestions of androgyny or unmanliness clustered around Davy for several reasons.[23] Simond acknowledged that the purple passages in his lectures had attracted ridicule. Displays of emotion could strengthen the affective bond with an audience and consolidate the message about the sublime passions chemistry was capable of arousing; but they could also undercut the lecturer's authority. An emotional lecturer might be viewed as one who had failed to master his own passions, hence tarnishing his image of manly self-control. At the same time, Davy's apparent willingness to accept honors and aristocratic patronage was seen by radicals and some reformers as diminishing his masculinity. Thomas Paine wrote that such subservience "marks a sort of foppery in the human character, which degrades it. It reduces man into the diminutive of man in things which are great, and the counterfeit of woman in things which are little."[24] Even Davy's attempt to mold his manners to pass in the upper-class circles he began to frequent could be portrayed as unmanly. As a social climber, he was routinely criticized for the artificiality of his deportment and dress. Even before he tried to fill Sir Joseph Banks's chair as president of the Royal Society in the 1820s, commentators were snidely remarking on the crudity of his clothing, gestures, and conversation. In 1813, the *Times* dubbed him *"dirty finger* gentry."[25] In trying to overcome this kind of prejudice, Davy ran the risk of appearing effeminate by devoting excessive attention to his dress or manners.

In addition, a lot of attention was devoted to his physiognomy, especially in his younger years. John Davy was concerned in his biography to answer allegations that his brother had presented a rustic and uncouth appearance on his first arrival at the Royal Institution. He quoted Davy's old friend Thomas Poole to the effect that "though his manners were retreating and modest, he was generally thought naturally graceful; and the upper part of his face was beautiful. I remember when he first lectured at the Royal Institution, the ladies said, 'Those eyes were made for something besides poring over crucibles.'"[26] Arguing that Humphry was in fact naturally elegant and well-mannered, John ended up making him sound quite feminine. Davy's physiognomy was described by his brother in remarkably feminine terms:

> I remember once a gentleman speaking to me about [Davy's hair],
> and expressing his admiration of its quality, very much in the manner

he might use in speaking of a lady's hair. His skin was delicate, and his complexion fair, with a good deal of colour. His countenance was very expressive, and responsive to the feelings of his mind; and when these were agreeable, it was eminently pleasing, I might say beautiful, for his smile was so.[27]

There is no reason to think Davy was in fact particularly effeminate in his manners or appearance. His rise up the social ladder presented him with a cruel dilemma: either retain the deportment and dress of his origins and be judged crude and provincial, or adapt to the new circumstances and be thought artificial and mannered. It is not surprising that he attracted some degree of ridicule by the way he chose to respond. Similarly, the eloquent passages in his lectures no doubt reflected his own poetic sensibility, nurtured by his early association with the romantic poets, rather than any specifically feminine trait of his mentality. The allegations of unmanliness are best interpreted not as pointers to Davy's essential personality but as symptoms of the tensions and anxieties surrounding gender relations in his culture. The strains in relations between the sexes were expressed in the newly available language of male and female psychology. Davy was accused of effeminacy, partly because of the significant social elevation he achieved in the course of his life and partly because of the appeal of his lectures to women and his endorsement of their scientific education.

Probably the most savage critique of Davy's character to appear in print was that launched by the *John Bull Magazine* in 1824. By this time Davy had retired from lecturing and was basking in the titles of baronet and president of the Royal Society. The anonymous author of the *John Bull* article, however, cut him down to size as number three in the series "Humbugs of the Age." While the author professed admiration for Davy's chemical discoveries and scientific reputation, he was sternly critical of the chemist's pretensions to gentility. The satire thus unfolded from the familiar Tory standpoint of condescension toward

Davy's lowly social origins and his all-too-obvious self-fashioning. The author gave his satirical thrust an extra twist, however, by labeling Davy a dandy. The chemist was said to have affected a "dandyish" dress—a green velvet waistcoat with gold spangles—that was unsuitable for a natural philosopher. He had adopted also the "dandyism" or "puppyism" of the intellectual who preens himself in mixed company: "The poor fellow fancies himself irresistible among the girls, and is evidently pluming himself, while conversing with them."[28] Far from contributing to a glamorous or amusing reputation, the writer asserted, this behavior succeeded only in making him look a fool.

The *John Bull* article went further, connecting Davy's "dandyism" with the disruption of normal gender roles in his milieu. For this disruption, Jane Davy was said to be largely responsible. As a noted Edinburgh bluestocking, she had flourished in the salons copied in the Scottish capital from the Parisian model. Stories were told about the devotion paid to her by male intellectuals in her coterie, where she was said to have presided over a set of degendered "old women, male and female." The *John Bull* writer thundered: "This mixture of dandyism and science, which has always appeared to us one of the most disgusting things in the world, gave the ton to the Edinburgh society, and Mrs. Ap. was up to her eyes in blue."[29] Such aspersions were commonly cast on Jane. Even the novelist Sir Walter Scott, her cousin and friend, complained that she had taken "the *blue* line" and committed herself "to rule the Cerulean atmosphere."[30] By marrying her, Davy was thought to have confirmed the predisposition to subject himself to uppity women that was already apparent in his Royal Institution lectures. In his cultivation of a female audience for chemistry, he had already connived in a challenge to male dominance. His submission to an intellectual woman in the context of marriage was thought to continue the pattern of encouraging female insubordination.

By designating Davy a "dandy," the *John Bull* author invoked the connotations of that term in the Regency period. Thomas Carlyle captured important features of the type in a famous chapter of his *Sartor Resartus* (1833–34). "A Dandy," Carlyle recorded, "is a Clothes-wearing

Man, a Man whose trade, office and existence consists in the wearing of Clothes."[31] As Ellen Moers showed in a classic study, the original dandy was somewhat more specific than the traditional "fop" or pretentious dresser. The archetypal dandy, Beau Brummell, sartorial and behavioral trendsetter of the Regency period, dressed particularly but not extravagantly: starched high collar, meticulously tied brilliant white cravat, unadorned blue or green coat, and plain buff breeches or pantaloons.[32] This was much the way that Davy was shown in Sir Thomas Lawrence's portrait, painted about the time of his accession to the presidency of the Royal Society in 1820 [Figure 5]. It was not clear that Davy fully matched the definition of a dandy because he worked as a chemist (often regarded as a menial occupation) and had only fairly recently escaped paid employment. But these characteristics only made the application of the label more fitting as a tool of satire. One of his obituaries declared that, in his dress, he "oscillated between a dandy and a sloven."[33] To call him a "dandy" was somewhat ironic, then, but not entirely inappropriate. As well as referring to ostentatious clothing, it invoked resonances other than the purely sartorial, for the dandy was recognized as a character type as well as a style of dress. As a personality, his defining feature was his dependence on the gaze of spectators for his sense of identity. Carlyle also caught this nuance well. The dandy, he noted, sought only "that you would recognize his existence; would admit him to be a living object; or even failing this, a visual object, or thing that will reflect rays of light . . . do but look at him, and he is contented." The dandy, as James Eli Adams noted, represents the ideal of "the hero as spectacle."[34]

Such a characterization could obviously draw upon Davy's self-presentation in the lecture theater. The charge that he was a man whose vanity required him to display himself for the admiration of spectators was also leveled, after his retirement from lecturing, in criticism by the *Examiner* and the *Times* of his wartime trip to Paris at Napoleon's invitation in 1813–14. The *Examiner* had no doubt as to Davy's motivation for crossing the Channel to enter enemy territory: "he may talk about so many chemical intentions as he pleases, but he goes to see and to be

Figure 5. Portrait of Humphry Davy by Sir Thomas Lawrence, 1821.

Painted shortly after his election as president of the Royal Society, this portrait of Humphry Davy was first exhibited at the Royal Academy in 1821. It shows Davy in fashionable Regency dress and adopting the swagger pose typical of Lawrence's masculine portraits. Lawrence and Davy shared similarly humble West Country origins (Lawrence was the son of a Bristol innkeeper), and they associated with each other while holding the offices of president (respectively) of the Royal Academy and the Royal Society. Both were also regarded as conforming, to some degree, to the stereotype of the dandy. Richard Holmes has written of Lawrence: "There remains a kind of mysterious doubleness about his identity, a secretive and possibly bisexual quality that runs both through his life and through his art" (R. Holmes, *Thomas Lawrence Portraits*, 10). See also Funnell, "Lawrence among Men"; R. Walker, *Regency Portraits*, 1:148–49. The original is held by the Royal Society, with copies in the collections of the Royal Institution and the National Portrait Gallery.

Portrait of Humphry Davy, RS 9343. © The Royal Society.

seen . . . to have it said, as he moves along through smiles of admiration and shrugs of obeisance, — 'Ah, there is the *grand philosophe, Davie!*'" Such conduct was particularly to be expected in France, which the English had long seen as the realm of effeminate courtliness. Napoleon's reestablishment of a court in Paris had resurrected the English stereotype of the ancien régime. The report concluded that Davy's behavior was both un-English and unphilosophical, and "such as goes hard to establish that charge of foppery which is made against Sir HUMPHREY's [*sic*] character in general."[35]

Some of Davy's supporters rallied to defend his venture into the enemy's domain. A correspondent wrote to the *Times* in October 1813 to insist on Davy's "powerful and manly mind." "Sir HUMPHREY DAVEY [*sic*] is not one of those vain and empty Charlatans, who must go all lengths to obtain the incense of popular flattery," this writer concluded.[36] But what was denied here was precisely what was being asserted elsewhere. Davy's susceptibility to female adulation was widely believed to have unmanned him, and his visit to France to soak up the praise of the Parisians was thought to provide all the confirmation that was needed. After his arrival in the French capital, stories of his dandyism continued to circulate. One anecdote, about his visit to the Louvre in 1813, survived to appear in J. A. Paris's biography, *The Life of Sir Humphry Davy* (1831):

> The English philosopher walked with a rapid step along the gallery, and, to the great astonishment and mortification of his friend and *cicerone*, did not direct his attention to a single painting; the only exclamation of surprise that escaped him was — "What an extraordinary collection of fine frames!" . . . They afterwards descended to a view of the statues in the lower apartments: here Davy displayed the same frigid indifference towards the higher works of art. . . . The apathy, the total want of feeling he betrayed on having his attention directed to the Apollo Belvedere, the Laocoön, and the Venus de Medicis, was as inexplicable as it was provoking; but an exclamation of the most vivid surprise escaped him at the sight of an Antinous, treated in the Egyp-

tian style, and sculptured in *Alabaster.*—"Gracious powers," said he, "what a beautiful stalactyte!"[37]

Predictably, perhaps, this was one of the passages from Paris's work to which John Davy took strong exception and which he pointedly omitted from his own biography of his brother. Interpreted in light of the critical press coverage of the trip at the time, the story seems to describe an instance of Davy's dandyism. Walking through the Louvre, Davy is there not to see but to be seen. As a dandy, his emotional satisfaction comes from self-display; he is unable to focus his affective energies on any external object, and hence is incapable of genuine aesthetic appreciation. In Davy's case, this idea is expressed through a boorish concentration on the technological (the frames) and the geologic (the stalactite) rather than on the central aesthetic attributes of the objects he confronts. One might defend Davy's conduct, as his brother did, by claiming that he was reluctant to admire such works as the Apollo Belvedere, the Laocoön, and the Venus de' Medici because he disapproved of their having been brought to Paris as loot from Napoleon's occupation of Italy.[38] It is possible that this was the intention of his behavior, but its meaning was scarcely transparent. Davy's contemporaries seem to have more readily interpreted his conduct as an example of the dandyism they believed typical of him, and it was likely for this reason that his brother tried to suppress the anecdote.

The characterization of Davy as a dandy had a further dimension, one connected with more intimate features of his sense of self. A dandy was generally supposed to be so consumed with narcissistic self-approbation that he was incapable of intimate relations with another person. In this respect also Beau Brummell was the ideal type. The dandy flirted with women but shunned involvement. Marriage and fatherhood were so antithetical to the role that a gentleman who wished to adopt it would have to conceal such attachments, if he was unfortunate enough to have them. The dandy was not typically identified as homosexual, but his capacity for heterosexual relations was seen as compromised or diminished. Investment of affective energies in self-display was thought to have left him unable to achieve fulfill-ment in the masculine sexual role.[39]

Again, the label captured certain features of Davy's public charac-ter dating back to the Bristol days. Although satirical attacks on the nitrous oxide episode had fastened upon the gas's potential to unleash heterosexual behavior, at least some members of the Bristol coterie seem to have been interested in strengthening the affective bonds be-tween men. The poets Southey and Coleridge, who breathed the gas under Davy's supervision, were both emotionally drawn to him. It is also possible that Gregory Watt was involved in some kind of homo-erotic relationship with another young man, William Creighton, an employee of Watt's father in Birmingham. The possibility is suggested by the scatological and sexual banter in Creighton's letters to Watt.[40] There is no evidence that Davy himself shared this interest. He was never accused of homosexual inclinations, even by his most savage critics, though it is just possible that the report that it was a statue of Antinous (the deified male lover of the emperor Hadrian) that caught his eye in the Louvre in 1813 was intended to spread innuendo along these lines. To make such an allegation explicitly would have caused considerable outrage, due to the heightened climate of homophobia in the period.[41]

In this climate, the affection that Coleridge in particular sustained for Davy over several years remained almost entirely private. The poet's "sympathy" and "love" for the young chemist was nonetheless intense. Coleridge taught himself chemistry, as he explained to Davy in 1801, "both for it's [sic] own sake, and in no small degree likewise, my beloved friend!—that I may be able to sympathize with *all*, that you do and think." He hoped his blind sympathy, "from the very middle of my heart's heart," would be illuminated by the light of shared knowl-edge, if only he knew more about the science. Speculating about the great discoveries the young Davy was destined to make, he asked, "To whom shall a young man utter *his Pride*, if not to a young man whom he loves?"[42] When Davy moved to London, Coleridge wrote to him frequently from his residence in the Lake District, bemoaning their separation and urging him to visit. In 1801–2, he attended a course of Davy's lectures at the Royal Institution, and he subsequently lectured there himself at Davy's suggestion.[43]

Unique in the intensity of his affection for his friend, Coleridge was in a privileged position to feel the moral and physical dangers of Davy's situation in London and to elaborate on them in striking language and imagery. In January 1804, he wrote to Southey describing a vivid and shocking dream, in which he encountered Davy reduced to "a wretched Dwarf with only three fingers." He was told that his friend, "in attempts to enlighten mankind had inflicted ghastly wounds on himself, & must henceforward live bed-ridden."[44] In this terrifying dream image, Coleridge conjured up Davy's mutilated body, a body that was glamorized and fetishized in the widespread commentary on his physical attractiveness. He was not the only person to be horrified by the prospect of such hazards assailing his young friend. In 1807, when Davy succumbed to a real, apparently life-threatening, illness, Coleridge reflected on the irony that it should have followed upon Davy's "attempts to enlighten mankind," just as in his dream.[45]

Coleridge also foreshadowed another trait of the commentary on Davy when he deliberated on less tangible threats stemming from the chemist's dedication to the general good. He mused about how Davy's devotion to the welfare of humankind could diminish his potential for personal erotic fulfillment. Clearly referring to Davy, though without naming him, he wrote to Southey in October 1801 that a chemist's investment of his affective energies in inanimate objects and in distant prospects of human benefit could damage his capacity to form an emotional bond with another person:

Yet I do agree with you that chemistry tends in it's [sic] present state to turn it's [sic] Priests into Sacrifices. One way, in which it does it . . . is this—it prevents or tends to prevent a young man from falling in love. We all have obscure feelings that must be connected with some thing or other—the Miser with a guinea—Lord Nelson with a blue Rib-

bon—Wordsworth's old Molly with her washing Tub—Wordsworth
with the Hills, Lakes, & Trees—all men are poets in their way, tho' for
the most part their ways are *damned bad ones*. Now Chemistry makes
a young man associate these feelings with inanimate objects—& that
without any moral revulsion, but on the contrary with complete self-
approbation—and his distant views of Benevolence, or his sense of
immediate beneficence, attach themselves either to Man as the whole
human Race, or to Man, as a sick man, as a painter, as a manufacturer,
&c—and in no way to man, as a Husband, Son, Brother, Daughter,
Wife, Friend, &c &c.[46]

Love, Coleridge went on to explain, was not invariably directed
toward another human being. The miser could be said to be "in love"
with his money in just the sense that a man might be said to love a
woman. The chemist, similarly, was in love with science. He diverted
his emotions into attempts to benefit humanity at large, at the cost
of his prospects for personal happiness. Thus, the priest of science
was turned into a sacrifice to his god. The only "salvation" was for the
"young chemist . . . [to fall] downright romantically in Love" with an
appropriate—human—object.[47]

It is not known whether Coleridge shared these thoughts with Davy
himself. But the young chemist was also musing, at this stage of his life,
about his prospects for erotic satisfaction. As we saw in the previous
chapter, he wrote to his Bristol friend John King about having succumbed to sexual temptations in London.[48] When he entertained the
Philadelphia chemist James Woodhouse on a visit to the city in 1802,
he recorded the American's encounter with "the nymphs of Drury
Lane & Covent Garden," with whom he may also have been acquainted
himself.[49] At the same time, he continued to profess a deep affection
for Anna Beddoes, with whom he exchanged letters. In one particularly touching letter of December 1804, Anna teasingly reproved Davy
for asking her to forget all about him. She could never do that, she told
him, and she did not believe he really wanted her to, in view of their
fond recollections of the time they had spent together.[50] Anna was,

however, an unattainable love object for Davy, as long as she was married to his former employer, and in any case she soon redirected her passions toward his friend Davies Giddy.[51] In private reflections, Davy considered how he could balance affection for others with what he felt was his essentially solitary nature. It is telling that, when he tried to sketch out the narrative form his life might take in his early notebooks, he included episodes of love but not marriage.[52] He could envision extending his feelings to another person in the throes of passion but did not seem to contemplate a permanent commitment.

For more than a decade after his arrival in London, bachelorhood remained his condition. His rather abrupt alliance at the age of thirty-five with the widowed Jane Apreece prompted further comments about his dandyism, along the lines that Coleridge had anticipated. The author of the *John Bull* article suggested that something other than love had been the motivating factor. When Davy proposed to Jane, the magazine writer suggested, "Her blue stockingism was delighted to the highest, and his ambition of shining among the fashionables instead of lecturing to them, also received its gratification."[53] The author could not resist pointing out how ridiculous was Davy's dedication to his new wife of his lectures on agricultural chemistry, "which, as the book chiefly treats on analysis of dung and other manures, was a well turned compliment." The gesture was symptomatic of his "nonsensical affectation of conjugality in the face of the public." But the facade of conjugal affection was a complete pretense, the author proposed, and one that was rapidly dropped as indifference and quarreling began to characterize the couple's relationship.

A great deal of gossip circulated about the Davys' marriage, and much of it was consistent with this picture of a couple whose affections had rapidly cooled, to be replaced by mutual hostility. On first acquaintance, Jane reported that Davy was pleasant and unaffected by "all the fashion & celebrity of admiration."[54] After the wedding, she seems to have quickly found out this was not true. Lady Charlotte Bury, a courtier who was acquainted with the pair, reported that "Sir Humphry, accustomed to adulation, seems to fall into surliness or dulness where he meets it not." She ascribed his boorish behavior in company to his "under breeding" and always trying to be the center of attention.[55] Sir Walter Scott, who had known Jane since her childhood, said the problem was one of incompatible tempers, which the couple did not even try to conceal. "They quarrel like cat and dog," he recorded.[56] A few months after the wedding, when Davy injured one of his eyes in an explosion, it was reported by the Earl of Dudley that stories were going around "that Lady D. scratched it in a moment of jealousy."[57] Michael Faraday, who accompanied the couple on their Continental tour in 1813–15, recorded his own experience of Jane's bad temper, haughtiness, and "evil disposition."[58] The author and wit Sydney Smith, who as a friend had urged Jane not to remarry, observed the breakdown of relations between the Davys. In 1816, he summoned a series of chemical metaphors, in a letter to Lady Holland that hinted at impotence along with temperamental incompatibility:

> The decomposition of Sir Humphry and Lady Davy is entertaining enough. I wonder what they quarrelled about. . . . I cannot conceive any third body interposing to alter their affinities. Perhaps he vaunted above truth the powers of Chemistry and persuaded her it had secrets which it does not possess, hence her disappointment.[59]

After two extended tours of the continent, the couple gave up traveling together. Davy wrote to his brother in October 1823 that it seemed to be the eternal lot of the philosopher since Socrates to be exiled from home by a shrewish wife.[60] Sir Benjamin Collins Brodie, a prominent surgeon who knew Davy through the Animal Chemistry Club established under the auspices of the Royal Society in 1808, concluded that the alliance was dictated by self-interest on both sides: Davy gained a share of his wife's wealth, and she gained social rank from his knighthood.[61] Other commentators concurred with this interpretation and deduced that love had little role in the partnership. There was apparently sufficient speculation about why the marriage remained childless for John Davy to feel the need to address it in his

Fragmentary Remains of his brother in 1858. He placed the blame on Lady Davy, invoking her "irritable frame and ailing body," though he failed to explain how such a weak physique could outlast her husband's by a quarter century. Jane's sickly constitution, and the couple's acceptance that it precluded the possibility of children, was, John insisted, "explanatory of much in the married life" of the Davys.[62]

By offering this explanation for the childlessness of his brother's marriage, John Davy tacitly admitted the existence of gossip less flattering to Humphry's manly reputation. John did the best he could in the face of rumors of Humphry's henpecked subordination to a woman of masculine character and ambitions. This is not to say that such was the unanimous view of the marriage. The *Annual Biography and Obituary* the year after Humphry Davy's death claimed that Jane had been "an affectionate and exemplary wife, and a congenial friend and companion" to her husband.[63] A few years later, David Brewster wrote in the *Edinburgh Review* that the partnership was based on mutual esteem, which "gradually ripened into affection."[64] The surviving letters between the two are not inconsistent with this supposition. Davy was an eager wooer, though the language in which he pressed his suit might seem rather formal to us. The letters following the marriage are cordial, if not particularly passionate. Davy wrote frequently to his wife during the Continental travels of his later years, reporting on his own state of health and inquiring solicitously after hers. He did not seem to fear her jealousy when he told her about his affection for a young woman who had nursed him during an episode of illness.[65] In the final crisis of his life, after he suffered a serious stroke in Rome in February 1829, Jane proved steadfastly loyal. She traveled halfway across Europe to accompany him on his last journey. After his death, she did what she could to memorialize his achievements, commissioning Dr. Paris to write the first biography using papers and letters she had collected. Her relations with Davy's brother were strained, however, and when Paris's biography turned out to be controversial and unsatisfactory, it was John Davy who assumed the guardianship of his brother's reputation.

Davy's marriage was not entirely visible to the public or even to close friends, and it would be unwise to rely on rumor and innuendo to try to establish its real character. But the prevalence of that gossip is itself interesting for what it reveals about his public reputation. Stories circulated about his shrewish wife and their loveless relationship because they were consistent with other things people thought they knew about Davy.[66] In particular, the gossip fitted with the image of him as a dandy: a man who subordinated himself to ambitious women in seeking their admiring gaze and who was so disabled by his dependence on being seen that he could not establish an intimate bond with another individual. The image was shared by some people who clearly resented Davy's success and some who claimed to admire him. It had even been foreshadowed in the reflections of someone who professed to love him and be deeply concerned about his welfare—namely, Coleridge. Coleridge had the advantage of closeness to Davy in his early years, but his remarks were echoed in their main themes by other contemporaries. Descriptions of Davy's clothing and deportment were made to bear the weight of moral disapproval of his personal self-fashioning. Coleridge's focus on the emotional displacements that followed from the chemist's role was also reproduced in other commentary. Portrayed as a dandy, Davy was thought to have been emotionally disabled by the admiring gaze of spectators. Surrounded by masculine women, and eventually submitting to one of them in marriage, he was thought to have crippled himself sexually. The supposition seems to have been in the background of at least a few of the comments on Davy's marriage.

Intriguingly, a kind of fetishistic obsession with Davy's body persisted well beyond his lifetime. In the 1930s, the Marxist journalist and historian J. G. Crowther reinscribed the detailed physiognomic descriptions given of him, repeating the very phrases used by contemporaries to depict his complexion, hair, eyes, and voice. The inappropriateness of his dress on occasion was noted and related to his "coxcomb" behavior and "pursuit of snobbery." Crowther even added speculations about the great chemist's physiology. Davy, he suggested, had "lived swiftly," at a high metabolic rate, which would explain why his significant discoveries were punctuated by periodic illnesses and terminated by an early death.[67] Even in the twentieth century, it seems, questions of Davy's moral character continued to be refracted through discussion of his physical embodiment. To that extent, his persona as a dandy was never quite laid to rest.

4. The Discoverer

Some miracle might have produced it, yet the stages
of the discovery were distinct and probable.
MARY SHELLEY, *Frankenstein*

Humphry Davy lived in an age of radical change in scientific disciplines and institutions. In the first couple of decades of the nineteenth century, entirely new disciplines—including biology, geology, and physiology—emerged. At the same time, existing ones, especially chemistry and physics, were fundamentally altered in their contents and methods. These changes went along with widespread reform of scientific institutions and the foundation of many new ones. New universities were established in Berlin and London. National and local academies were reorganized during the revolutionary and Napoleonic eras in France. In Britain, new learned societies were formed in many towns and cities, and national organizations, including the Royal Institution, took on the duty of communicating scientific information to the general public. To some, the new institutions suggested that scientific discovery itself might be reduced to mechanical routine. In a famous essay of 1829, Thomas Carlyle conjured the image of a scientific genius replaced by the workings of social machinery:

No Newton, by silent meditation, now discovers the system of the world from the falling of an apple; but some quite other than Newton stands in his Museum, his Scientific Institution, and behind whole batteries of retorts, digesters and galvanic piles imperatively "interrogates Nature,"—who, however, shows no haste to answer.[1]

Davy may well have been Carlyle's target here, but, if so, this was another criticism the man himself would have rejected. He did not regard machinery as antithetical to the expression of human genius. On the contrary, he appropriated electrical apparatus as a personal attribute, an accessory to his individual identity [Figure 6]. In the theater and the laboratory, he deployed the electrical battery as a prosthetic extension of his embodied genius. In disciplinary terms, it consolidated his position within the field of chemistry, the scientific domain with which he was most closely identified. The link between the two was his persona as a *discoverer*. Davy used the power of electricity to discover several new metallic elements. Produced by electrolytic analysis, they provided new building blocks for the composition of matter, and they placed him at the forefront of what was sometimes called the Augustan Age of chemistry. David Brewster wrote that Davy's discoveries had outshone all others since the age of Newton.[2] It was taken as a sign of his genius that he had realized the potential of voltaic electricity to tear apart substances and reveal their constituent elements. He had harnessed a power unknown to previous chemists, including Antoine Lavoisier, who had initiated the so-called chemical revolution with his new theory of combustion in the 1770s but whose grasp of composition was now shown to be incomplete.

Dramatic new discoveries were hailed as foundational for several new or reconstructed scientific disciplines at this time, as Simon Schaffer has noted.[3] Discoveries, whether in chemistry, astronomy, or electricity, were often associated with individuals who were identified as geniuses in line with the romantic stereotype. Such discoveries were thought to be signs of the superhuman insight that characterized genius. Historical scrutiny, however, usually discloses a more complex picture. On closer examination, we find that events were categorized as discoveries and assigned to particular individuals through sometimes

Figure 6. Portrait of Humphry Davy by Archer James Oliver, 1812.

Less famous than the Howard or Lawrence portraits, this painting presents a less glamorous image of Humphry Davy. His facial features and expression appear much less attractive than in other depictions. He is shown as if at work, surrounded by the apparatus with which he made his great discoveries at the Royal Institution, including a tray from the voltaic battery on the floor to the right. Exhibited at the Royal Academy in 1812, the painting was donated to the Royal Institution in 1889. See Prescott, "Forging Identity," 73; R. Walker, *Regency Portraits*, 1:149.

Royal Institution of Great Britain / Science Photo Library (C004/6725).

lengthy and involved processes. The relations between the discoverers and the disciplines to which they supposedly gave rise were also more complex than has often been assumed. Davy's case provides a good example of how a person came to be assigned the attributes of genius and authorship of several significant discoveries. We will see that his authorship was not uncontested, that in fact Davy had to struggle to secure credit for his discoveries and to beat off challenges to them. We will also see how troubled was Davy's relationship to the discipline of chemistry. Far from having spontaneously originated a new kind of chemistry, his efforts to reform the discipline on the basis of his findings were largely unsuccessful. Even after securing his credentials as a discoverer, he found he could not give birth to a new discipline from his own powers of genius alone.

Davy made his discoveries by recruiting voltaic electricity to unlock the secrets of chemical composition. He used the instruments Carlyle was referring to when he talked of "batteries" and "galvanic piles" to expand the range of phenomena that were considered part of chemistry. He showed that electricity was the force that held matter together and provided the means to prize it apart. But Carlyle was wrong to suggest that, in Davy's day, this had already become an institutionally routinized accomplishment. On the contrary, Davy's discoveries were presented and secured by his charismatic authority, articulated in bodily performances in the theater and the laboratory, and translated more or less effectively into such writings as lectures and research reports.[4] By these means, he came to be recognized as the leading chemist of his time, and his findings of new elements were widely accepted. On the other hand, his influence was never translated into a regime for training students or a satisfactory textbook. It remained tied to Davy's embodied persona and his authorship of certain kinds of writings. In the longer term, as Carlyle had foreseen, laboratory training regimes arose, and new pedagogical syntheses emerged alongside them. When that happened, Davy's precise role in the discipline of chemistry became hard to discern. He was remembered as a great discoverer, with several elements to his name, but he had evidently not succeeded in establishing a renewed discipline on those foundations.

Davy's interest in galvanism was aroused by Alessandro Volta's discovery of his "pile," or battery, announced in a letter to Sir Joseph Banks in March 1800 and soon made known to the Royal Society. Volta showed that an arrangement of alternating plates of different metals, interspersed with cardboard disks soaked in water or dilute acid, could generate shocks, sparks, and muscular spasms similar to those caused by static electricity. He made the news known against the background of Luigi Galvani's finding, in 1791, that a probe composed of two metals, when used to connect the muscles in a frog's leg to nerves in its spinal cord, caused motion in the leg. Galvani's finding had generated an intense and widespread controversy as to whether the effect was caused by animal electricity from the frog's nerves or by a physical interaction between the metals of the probe. Volta believed his discovery could resolve the controversy in favor of an inorganic cause for the impulse since he claimed to have shown that an arrangement of nonliving materials could produce the same effects. The debate was not concluded as readily as Volta hoped, but his apparatus was quickly replicated in many locations, and it set investigators on entirely new trails of inquiry.[5] Within a few months, the battery was used in London by the scientific journalist William Nicholson and the surgeon Anthony Carlisle for a dramatic experiment in which its current was passed through water. The outcome was the production of gases, hydrogen and oxygen, from the wires leading from the two poles of the battery. The result showed, according to Nicholson and Carlisle, that water could be decomposed into its component elements by voltaic electricity. The implication was that the pile could be used as an instrument of chemical analysis, perhaps one that would be capable of breaking up bodies that had resisted previous analytic methods.

This was the vision Davy was to realize in his work. His brother, John, wrote later that as soon as Humphry heard of Nicholson and Carlisle's experiment, Humphry "had prophetic warnings that it was a passage to a new world of discovery."[6] Yet there was a degree of mythologizing in this retrospective view. The path to use of the voltaic pile as an instrument of analysis was not as straight or smooth as it came to seem in hindsight. Even with his renowned powers of genius, Davy could not foretell exactly what the capabilities of Volta's device would be. At first, he interpreted the voltaic pile in connection with his interest in the processes of life and the role of so-called imponderables. He heard about the invention as he was concluding his investigation of nitrous oxide with Thomas Beddoes in Bristol, immediately constructed a pile, and set about experimenting with it. Like nitrous oxide, which had proved "capable of producing the greatest changes in the phaenomena of life," galvanism seemed to point toward the "sublime chemistry" that would reveal the processes underlying living things.[7] The voltaic pile promised further disclosures of life's secrets. When he first heard about Nicholson and Carlisle's result, Davy wrote to his friend Davies Giddy, "An immense field of investigation seems opened by this discovery: may it be pursued so as to acquaint us with some of the laws of life!"[8] A few months later, he reported again to Giddy on his experiments. Davy had established, he claimed, that galvanism "is a process purely chemical & it depends wholly on the oxydation of metallic surfaces having different degrees of electric conducting power." He expressed regret that he had no room in his letter to "expatiate on the connection which is now obvious between galvanism & some of [the] phaenomena of organic motions."[9] The following month, he boasted to his poet friend Samuel Taylor Coleridge that he had "made some important galvanic discoveries which seem to lead to the door of the temple of the mysterious god of Life."[10]

These hopes spurred Davy in his investigation of the voltaic pile, reported in several published letters in Nicholson's *Journal of Natural Philosophy, Chemistry and the Arts* in the second half of 1800.[11] In a series of experiments, which varied the composition of the pile and applied its current to different objects, Davy strengthened his conviction that its effects were fundamentally chemical in origin. He fastened upon the oxidation of the metals in the pile as the basic cause of the current. At the same time, however, he refused to segregate galvanic phenomena from their physiological manifestations. He constantly used his own body to measure the strength of the current. He graded shocks in terms of their painfulness and how much of the body they affected.[12] He also passed the current through samples of dead muscle fiber. As he experimented, his speculative musings, both public and private, led in several directions. On the one hand, Davy was convinced that Volta's instrument provided a key to explaining the cause of Galvani's phenomenon. The pile produced electricity by a chemical process, and muscles and tissues were simply making the effect manifest. On the other hand, Davy resisted following Volta to his ultimate conclusion: that animal electricity could be entirely reduced to inorganic reactions. In his final letter to Nicholson, written in January 1801, Davy declared that "there exists in living matter galvanic action independent of all influence generated by metallic oxydation."[13] Perhaps, then, the workings of the pile were not to be taken as a complete model for galvanic processes in living things? Davy did not want to give up the idea that other causes, perhaps involving imponderable or ethereal entities, were in play.

Davy's speculations at this time also continued a line of thought concerning the role of light and other imponderable fluids in biological processes, on which he had embarked in his essay in Beddoes's volume *Contributions to Physical and Medical Knowledge* in 1799. He suggested that galvanism and electricity—at this stage not definitively identified with each other—were additional imponderables that must be related to the nervous fluid and thus to animal motion.[14] Alluding to this idea, Robert Southey wrote to him in July 1800 that "it should appear as the galvanic fluid . . . is the same as the nervous fluid, & your systems will prove true at last."[15] In a lecture given shortly after his arrival at the Royal Institution, in September 1801, Davy declared that the voltaic pile "cannot fail to elucidate the philosophy of the imponderable or ethereal fluids."[16] As Giuliano Pancaldi has noted, Davy even invoked the theory of phlogiston (the supposed principle of inflammability that was often regarded as another imponderable fluid) in private speculations about the interpretation of Nicholson and Carlisle's crucial experiment.[17] Although it had served as inspiration for his own inquiries, Davy was not convinced at first that the experiment demonstrated the analysis of water into its constituent parts, as its authors claimed. Though he kept these doubts out of his published letters to Nicholson, he wrote of them to James Watt, who he knew harbored sympathies for the old phlogiston theory, even though it had supposedly been disproved by Lavoisier's discovery of the composition of water fifteen years earlier. In a letter to Watt in January 1801, Davy wrote that the hydrogen produced at one of the wires from the battery was perhaps formed by combination between an "invisible" substance and a "quantity of the gravitating matter of water." He asked rhetorically, "May not this substance come from the metal & be the old phlogiston?"[18] In another letter at the end of December that year, he wrote again, promising Watt that he would soon tell him about further galvanic experiments, which "I [am] almost sorry to say cannot be well reconciled to the French theory & I even begin to believe that your theory of water being the basis of all the gases will be found true at last."[19]

Sharing these thoughts with Watt, Davy was keeping open a line of communication with the Lunar Society of Birmingham, where ideas about imponderable fluids had flourished in the previous generation. Joseph Priestley, a leading advocate of these ideas, was writing from exile in Pennsylvania about his own experiments with the voltaic pile, which he believed upheld the phlogiston theory.[20] Davy continued to speculate, from time to time, about the possibility that some version of pre-Lavoisian chemistry might be revived. He even mentioned it in a footnote in his Bakerian Lecture to the Royal Society in 1807, where he noted that a kind of phlogiston theory could be defended by supposing that all metals were compounds of certain bases with hydrogen, and oxides compounds of the same bases with water.[21] He allowed that this would be less elegant than Lavoisier's theory because it would require the hypothesis that additional (currently unknown) substances existed. But he still continued to toy with the idea. In 1809, James Dinwiddie heard him say that he thought a modified version of the phlogiston theory was still tenable.[22]

Fairly soon after arriving at the Royal Institution, however, Davy swung decisively against the variant interpretations of Nicholson and Carlisle's experiment. He realized that using the voltaic pile as an instrument of analysis required one to accept that it acted to separate water into its constituents, hydrogen and oxygen. As he began to exploit its potential as a tool of discovery, he came to treat it in an instrumental manner, putting aside fundamental explorations of its mode of action and detaching it from the context of physiological investigation. There were several reasons for his move in this direction. In his "Discourse Introductory to a Course of Lectures on Chemistry," delivered in 1802, he emphasized the powers conferred upon a discoverer by the effectiveness of the instruments under his command. He clearly saw the voltaic pile as the most important such instrument available to him. The following year, an apparatus of one hundred double plates was constructed under his supervision for his use at the Royal Institution. Turning away from physiological researches, Davy was directed to employ the machine in a series of assays of metallic substances, including the newly discovered palladium. As a result, he became accustomed to using the device as an analytic tool, and — at least in public — largely abandoned speculations about its relation to imponderables and the processes of life.

An additional consideration, as Iwan Morus has suggested, may well have been Davy's desire to distinguish himself from those offering less respectable forms of scientific performance in London.[23] Several practitioners were using galvanism in dramatic, if somewhat gruesome, public displays. Dinwiddie saw the lecturer John Tatum apply the current from a voltaic pile to a decapitated hare and a dead frog. He was appalled at Tatum's assertion that he had succeeded in restoring the frog to life, when — according to Dinwiddie — all that had happened was that he had elicited convulsive spasms in its muscles.[24] The most notorious shows of this kind were mounted by Giovanni Aldini, a nephew of Galvani, who visited London in 1802–3. Aldini's demonstrations of the power of galvanism to cause motions in dead animal tissue climaxed in experiments on the body of a hanged criminal, George Forster, in January 1803. In the presence of leading surgeons and press reporters, he succeeded in producing some movement of the limbs of the corpse, and one of its eyes was said to have opened. For Aldini, the results suggested that galvanism offered hope of restoring vital functions to the bodies of the recently deceased.[25]

Aldini's displays caused a sensation, but few observers were convinced by his general arguments. In October 1803, an anonymous review of his *Account of the Late Improvements in Galvanism* appeared in the *Edinburgh Review*. Authorship of the notice has been ascribed to Davy, though it seems unlikely that he actually wrote it.[26] Whoever he or she was, the reviewer was scornful of Aldini's claims, which were said to show "the vanity of systematizing upon an imperfect series of experiments." The writer disputed the notion that animal tissues were capable of producing their own distinct kind of electricity, insisting that the phenomena referred to by Aldini could be explained on the theory that the galvanic current was caused by chemical reactions between metals. As we have seen, this was Davy's preferred theory of the voltaic pile. As for Aldini's experiments on the bodies of the dead, the reviewer judged them to be "rather disgusting than instructive."[27]

Whether or not he wrote this review, Davy clearly intended his performances before the respectable audiences at the Royal Institution to be different from shows like Aldini's. As we have seen, he deployed the institution's voltaic pile as an integral part of his self-presentation in the theater, showing his audiences how the bodily senses could be used to detect and measure its effects. Touch, smell, sight, taste, and hearing were all invoked to assess the currents and sparks produced by the apparatus. This method of presentation also confirmed Davy's persona as an individual who deployed his refined sensibility in experimental work, as he wrote of the pile that "the most delicate organs are the best fitted for performing and observing its operations."[28] In 1802, his "Discourse Introductory" paid rhetorical tribute to the power of such instruments, by which humankind was learning not merely to study nature but to master it. When a larger battery was made in 1808, Davy again presented it to his audience, who included the patrons who had funded its construction. Occasional mishaps in demonstrations of the pile—for example, one in June 1810, witnessed by Dinwiddie and reported in the press, when noxious fumes drove away some of the audience—were significant embarrassments.[29] Davy sought to mount smooth displays of the instrumental working of the pile, in order to translate into the laboratory the authority he accrued in the theater.

Securing this authority was a struggle for Davy in the first few years of the nineteenth century because there was no immediate consensus on the interpretation of Nicholson and Carlisle's experiment. The new periodicals of the time, especially Nicholson's *Journal* and Alexander Tilloch's *Philosophical Magazine*, gave space to a large number of experimenters who used their own equipment and advanced their own interpretations of what happened when the current from the voltaic pile was passed through water. Some, including Priestley in March 1802, seized on the experiment as support for the phlogiston theory, along the lines of Davy's own conjecture in his letter to Watt the previous year.[30] Others, including the German natural philosopher Johann Wilhelm Ritter, took the fact that hydrogen and oxygen were produced at separate wires leading from the two poles as proof that the gases could *not* be the component parts of water.[31] What must be happening, on this account, was that one kind of electricity was combining with water to yield oxygen, and another kind was combining to yield hydrogen. A further complication arose in 1805, when investigators from England and from Italy reported that they had used the voltaic pile to produce acids from water. These acids could not be products of analysis but again would have to be seen as the results of combinations of electricity with water or components of it. If these reports were to be accepted, it would imply that the voltaic pile was not an instrument of analysis, or at least not one that could be used reliably for that purpose.

Davy confronted this confusing situation in his first Bakerian Lecture, "On Some Chemical Agencies of Electricity," delivered to the Royal Society on 20 November 1806. His achievement in that lecture was to demonstrate that, with proper precautions and procedures, the voltaic pile could be used as an effective instrument of analysis. He did so by reaffirming Nicholson and Carlisle's claim that it acted simply to separate water into its component gases, hydrogen and oxygen. The foundations of the argument were experimental, its style rhetorical, and its implications sociological. Experimentally, Davy introduced several refinements of technique that relied on the resources available to him in the laboratory of the Royal Institution. He made use of the battery built in 1803, attached expensive platinum wires to it, and dipped them into cups made of agate and gold and connected by a piece of asbestos. With meticulously purified materials in various arrangements, he was able to demonstrate that all the variant products claimed by other experimenters were due to contamination. Cups of resin, wax, or glass, or insufficiently purified water, could all harbor the contaminants that yielded the extraneous products. If they were eliminated, the pile could then be trusted to produce only hydrogen and oxygen, the true constituents of water.[32]

Rhetorically, the lecture was a masterful exercise in dialectical demonstration. Davy systematically closed off potential sources of contamination, carefully eliminating them one by one. At the same time, he seemed to anticipate and answer the objections of a skeptical reader, sustaining a kind of implicit dialogue in which the reader's doubts were addressed before they could even be articulated. Several reviewers attested to the rhetorical effectiveness of the lecture. The normally cynical Henry Brougham in the *Edinburgh Review* noted that he felt "an irresistible disposition to confide in . . . the author," even in places where Davy had not described his reasoning in detail.[33] The forcefulness of the lecture also enhanced its sociological effect. Davy raised the bar for those who wished to be credited as users of the voltaic pile. Investigators who were working with homemade and fairly cheaply produced apparatus were put on notice that their contributions would no longer carry as much weight as those emerging from an establishment such as the Royal Institution. Henceforth, a higher level of experimental skill and greater material resources would be expected. Davy had raised the stakes for anyone else who wished to be taken seriously in this field of inquiry.

In a way, this was a maneuver of disciplinary consolidation. Davy's authority over the field of electrochemistry was further enhanced by his announcement of the discovery of two new elements, sodium and potassium, in the next Bakerian Lecture in November 1807.[34] The production of these metals, by voltaic decomposition of fused samples of their oxides, was offered as a vindication of his meticulous experimental methods and his command of the pile as an instrument of analysis. His claim that the apparatus could be used to take apart previously un-analyzed substances had borne triumphant fruit. In June of the follow-ing year, in his third Bakerian Lecture, Davy was able to bring forward four more new metals, isolated from alkaline earths by the same appa-ratus: barium, strontium, calcium, and "magnium" (later renamed magnesium).[35] Each discovery further bolstered his reputation as the leading champion of chemical analysis, wielding the powers of voltaic electricity to probe the composition of matter more profoundly than had ever been done before. It is no surprise to find him telling his Royal Institution audience in March 1808 that the voltaic battery had proved to be "the most wonderful and important electrical instrument . . . the most powerful agent that has yet been discovered for effecting decom-positions and new combinations."[36]

Davy's position was strengthened by his discoveries, but it was not uncontested. Brougham, notwithstanding his praise for the first Ba-kerian Lecture, disparaged the second one, finding "nothing which de-serves the name of genius in the whole investigation." He judged that anyone who possessed "the excellent apparatus of the Royal Institution, could have almost ensured himself a plentiful harvest of discovery."[37] The charge was a painful one because it suggested that Davy's much-touted genius was simply the result of diligent labor with the expen-sive resources conferred on him by his patrons. It hinted that Davy was monopolizing the tools of chemical science, giving himself an unfair advantage over other chemists. Davy was understandably sensitive to these allegations, which found an echo in Carlyle's comments two de-cades later. During his third Bakerian Lecture, Davy made sure to tell his audience that he was still using the battery constructed in 1803, even though it was now showing loss of power due to corrosion. He men-tioned this so that other chemists would not be deterred from pursuing similar experiments "under the idea that a very powerful combination was required."[38] When John Davy edited this lecture for inclusion in Humphry Davy's *Collected Works*, after his brother's death, he added a footnote to draw attention to this comment. He wanted the point to be appreciated by those "historians of science" who had incorrectly "attributed the author's brilliant success in electrochemical research to his supposed extraordinary means, the enormous Voltaic batteries of the Royal Institution."[39] Potent instrumental resources such as the battery were counted among the attributes of Davy's genius, but too much emphasis on their power could diminish his intellectual brilliance and highlight the material advantages of his institutional position.

Whatever his criticisms, Brougham did not cast doubt on the basic validity of Davy's findings. Such doubts were, however, voiced from other provincial locations. Charles Sylvester, an entrepreneurial chemist and inventor living in Derby, challenged the elemental status of sodium and potassium in his *Elementary Treatise of Chemistry* in 1809. Sylvester suggested that the supposed elements could actually be compounds of soda and potash with hydrogen.[40] More obscure individuals denied Davy's claims on an even more fundamental level. Ezekiel Walker, from Lynn in Norfolk, continued to dispute the decomposition of water by the voltaic pile and to uphold the phlogiston theory, in articles published through the 1810s.[41] George Smith Gibbes, a physician from Bath, took a similar line in a book published in 1809.[42] Davy disdained to reply to these criticisms, perhaps feeling that he could afford to ignore objections from such marginal individuals. Better connected and better resourced metropolitan investigators had generally fallen into line behind his discoveries. William Hasledine Pepys, an instrument maker and supporter of the Royal Institution, and John George Children, a Cambridge-educated gentleman of independent means, both built batteries comparable to Davy's and joined him in subsequent investigations.[43] At the same time, more formidable rivals were emerging across the Channel in the persons of the leading French chemists, Joseph Louis Gay-Lussac and Louis-Jacques Thenard, who were armed since 1807 with a voltaic battery of unprecedented size built on Napoleon's orders.[44] It was this threat that spurred the Royal Institution to mount a subscription among its supporters and to construct the two thousand-plate apparatus that Davy had at his disposal from 1808 onward.

By the end of the first decade of the nineteenth century, Davy had placed himself in a position of preeminence in electrochemistry. He had assumed leadership of a national effort, directed at defeating the rival program based in Paris. As his brother later remarked, Davy had taken the part of a general, maneuvering his big battery on the field of scientific combat as the real generals of Britain and France were doing in the real battles between the two nations. In this situation, he could ignore the cavils and sniping of marginal figures in the provinces. What mattered was the support of his metropolitan audience, his elite patrons, and other expert chemists. His spectacular production of new elements had confirmed his reputation as a discoverer in their eyes, and he continued to deploy their support as needed. For example, he cited distinguished British chemists as witnesses to his experiments in the ongoing disputes with Gay-Lussac and Thenard, which indicates that his authority was still, in certain respects, localized. Although he had expanded his range of influence with his Bakerian Lectures to the Royal Society, published in that body's venerable journal, the *Philosophical Transactions*, his base of operations remained at the Royal Institution. He had built his reputation with the audience in the lecture theater and worked out how to translate it into support for his experiments downstairs in the laboratory. His persona as a discoverer was primarily built upon his work in the Albemarle Street building.

Davy's situation at this point had its strengths but also its weaknesses. Both were exposed in the course of the lengthy controversy that followed his announcement of the elemental status of chlorine in 1810. In two publications that year, Davy staked his reputation on the claim that what was usually called "oxymuriatic acid" was not an oxide of the familiar muriatic acid but rather an elemental substance that could not be analyzed into more fundamental components.[45] In this instance, unlike his previous discoveries, he had no new substance to exhibit. There were no new metals to flourish dramatically in the lecture theater. What he described was a series of experiments in which he had tried, but failed, to decompose the familiar greenish gas by the most powerful available voltaic battery and by other means. He also suggested how chemical theory could accommodate a new element of this kind, one that would be analogous to oxygen: an electronegative substance capable of combining with metals and combustible materials. In several ways, Davy relied upon his status in the chemical community to advance his claim. He invoked the definition of an element given by Lavoisier—any substance that could not be decomposed by available methods of analysis—and claimed that he himself, more than anyone else, knew what could and could not be accomplished by the available analytic instruments. Davy cited witnesses to the experiments he was adducing in support of the claim that oxymuriatic acid could not be decomposed. He asserted that if he could not break it apart, nobody could. And he assumed the prerogative of giving the purported element a new name, "chlorine," which he said had been approved by "some of the most eminent chemical philosophers in the country."[46]

Davy was not claiming to have discovered chlorine, just to have ascertained its elemental status. He insisted he was reporting factual results of experiments, not speculative interpretations. His assertion nonetheless encountered serious and sustained resistance, especially from John Murray, a private lecturer on chemistry and pharmacy in Edinburgh and the author of two popular textbooks. Murray mounted his own series of experiments to show that the gas in question *did* in fact contain oxygen. The controversy unfolded in the pages of Nicholson's *Journal* between February 1811 and April 1813 as well as in competing experimental displays in London and Edinburgh. I have described the dispute at length elsewhere, so there is no need to give further details here.[47] It is just worth pointing out how it reflected both Davy's authority within the chemical community at the time and the limitations on his influence. Davy mobilized all of his customary resources to defeat Murray's challenge. He staged demonstrations in the Royal Institution laboratory, where distinguished witnesses were called in to observe the crucial experiments, and in the lecture theater, where his audience was invited to confirm his interpretation of the results. Dinwiddie and the young Michael Faraday were among those who witnessed these performances, at which the elemental status of chlorine was supposedly manifested. Davy also aroused the patriotism of his audience by telling them that the discovery of chlorine had exposed the errors of the French system of chemistry, in which oxygen was considered necessary to the composition of acids.[48] He even got his brother, John, then a medical student in Edinburgh, to stage a demonstration in support of his claims in Murray's own city.[49] In this way, Humphry Davy capitalized on the authority he had built up through public displays in his own institution and used a surrogate to extend it geographically into a new location.

Davy's prestige was not accepted without question, however, and his writ did not run everywhere. His authority remained bound in certain respects to the locations where he had built his reputation. Murray was a resourceful opponent who shared some of Davy's skills as a public lecturer and experimenter. And he could command an audience in Edinburgh, the British city that most closely rivaled London as a center of intellectual and scientific activity. The dispute had several twists and turns, including the production of a new gas (subsequently named "phosgene") in one of John Davy's experiments, which the brothers claimed had been present in Murray's too and had led him astray. The controversy remained unresolved in early 1813, when John Davy curtailed discussion by simply refusing to reply to any further objections from his Scottish opponent. To judge from the statements of specialist chemists and the treatment of the topic in textbooks, the elemental status of chlorine was not generally accepted until around five years after this date. Humphry Davy's own attempt at a textbook, the *Elements of Chemical Philosophy* (1812), tried to show how chemical theory could be reconstructed around the assumption that chlorine was an element. But, as we shall see in the next section, this idea was resisted, and the whole project of the textbook—an attempt to recast the discipline from Davy's own point of view—was commonly regarded as a failure.

The identification of chlorine as an element was the culmination of Davy's work in chemical analysis. It came at the climax of his series of discoveries with the voltaic pile. He had systematically established the efficacy of the apparatus as an analytic instrument, detaching it from the context of physiological research and attempts to reanimate dead bodies, and then deploying it spectacularly to produce several previously unknown substances. As we have seen, these achievements relied upon the resources yielded by his specific location: the audience at his lectures, the patrons who supported his work, and the material assets of the Royal Institution laboratory. With this support, Davy was able to present himself as a discoverer and claim a position as the leading chemist of his day. His genius was widely recognized, and he was able to ignore the few individuals from marginal locations who disputed his findings. But, as the controversy with Murray indicated, his reputation was still strongest at its site of origin. When his published claims were challenged, he resorted to the support of his London audiences. His authority was concentrated geographically, and it was also limited in certain respects within the discipline of chemistry. By isolating a number of new elements, Davy had provided the foundations for a new structure of chemical theory. But he had not himself erected the new superstructure on the foundations he had laid. When he tried to do so, in his 1812 textbook, he encountered serious problems that led him to abort the whole project.

There had been textbooks of chemistry for more than two centuries by the time Davy embarked on writing one. They appeared on the scene when the discipline began to be taught in universities in the early seventeenth century.[50] The needs of pedagogy created a demand for books that would summarize what was known about substances and reactions and organize it in a systematic form. But such works did not always keep up with the changing frontiers of knowledge. In the late eighteenth century, chemistry was an enterprise with fluid boundaries and shifting foci of interest. Chemists' attention was drawn to new instrumentation and phenomena when pneumatic chemistry emerged in the 1760s and 1770s. The pioneers of that field, Joseph Black, Henry Cavendish, and Joseph Priestley, developed new techniques to produce gases from chemical reactions and learned how to distinguish their properties. But there was no immediate consensus about the implications of their findings, which did not get into the textbooks for several years. Though Black taught chemistry regularly, and Priestley occasionally did so, neither they nor Cavendish ever published a comprehensive textbook of the subject.[51]

The chemical revolution initiated by Antoine Lavoisier established a new pattern, since his program of disciplinary reform included a prominent pedagogical dimension. Lavoisier introduced new instrumentation of his own for handling gases and measuring the exchanges of heat in reactions. His oxygen theory provided an alternative to the traditional theory of phlogiston by way of explaining calcination and combustion. Oxygen was also proposed as the principle of acidity in chemical substances. Lavoisier recruited other leading French chemists to his cause in the 1780s and made two crucial contributions to the teaching of the subject. In 1787, along with three coauthors, he published a system of nomenclature, in which the new theories of composition were built into a language for naming chemical substances. Two years later, his *Traité Élémentaire de Chimie* (*Elementary Treatise of Chemistry*) provided a comprehensive textbook, in which the new theoretical framework was to serve as the basis for teaching students. However, Lavoisier had virtually nothing to say about the theory of affinities, the doctrine of the different strengths of attraction between different substances. Tables in which the relative strengths of affinities had been listed for easy memorization and reference had been a crucial part of chemical didactics for several decades. Lavoisier omitted them entirely from his *Traité*, thereby depriving himself of a standard pedagogical technique.[52]

Despite Lavoisier's efforts, when Davy was teaching himself chemistry in the 1790s, the theoretical picture was far from clear. It seemed that the French chemist had not bequeathed a fully formed new discipline. Despite his adoption of the rhetoric of "revolution"—a cruel irony in view of his execution by the revolutionary regime in 1794— Lavoisier did not instantly and comprehensively transform chemical theory. Many other chemists, in France and elsewhere, were respon-

sible for circulating and adapting his new theories, and institutionaliz-
ing them in pedagogical programs, in the years after his death.[53] Some
of Lavoisier's ideas, especially his theory of acidity, fell by the way-
side during this time. Others were reinterpreted and integrated with
new discoveries. Much of this has been overlooked by those who have
emphasized the revolutionary insights of a single individual, especially some twentieth-century historians who were looking for instan-
taneous gestalt switches or paradigm shifts in the history of science.
Such a perspective does not help us understand the situation in the
immediate aftermath of Lavoisier's career, when Davy came upon the
scene.[54]

As he began to familiarize himself with chemistry, the young Davy
studied Lavoisier's works alongside those of William Nicholson. He
therefore learned the new French theories in the context of their rather
skeptical appraisal by British chemists. Nicholson had cast doubt on the
French pretension to a high degree of accuracy in the reported measurements of quantities in reactions. Along with many British writers,
he resented the new nomenclature, which he saw as the imposition
of a theoretical interpretation on what should be a neutral language
of description. In successive editions through the mid-1790s, Nicholson's textbook presented Lavoisier's theory alongside the traditional
theory of phlogiston, inviting students to make up their own minds
about which was more plausible.[55] Reading Nicholson's *First Principles
of Chemistry* in any edition up to the third (1796), Davy would have
learned that Lavoisier's ideas had not won universal acceptance, which
no doubt influenced his own approach of selective appropriation and
criticism. In his 1799 "Essay on Heat, Light, and the Combinations of
Light," Davy rejected the material theory of heat, upheld by Lavoisier
(who called it "caloric"). Davy favored the alternative view—advanced
by the man who was soon to become his patron, Count Rumford—that
heat was a type of motion of the particles of matter. As we have seen,
he continued to raise the prospect of a restoration of some version of
the phlogiston theory at various points in his career. And his claim for
the elemental status of chlorine was accompanied by a direct attack on

Lavoisier's theory of the role of oxygen in acidity. For Davy, it never seems to have been the case that Lavoisier's "revolution" had to be accepted as a whole; he always saw it as open to modification.

By the time he came to write a textbook of his own, Davy had firmly established his identity as a discoverer of new chemical substances. He carried this identity into his authorship of the *Elements of Chemical Philosophy* (1812), of which part 1 of volume 1 was the only portion to appear. Although formally a textbook, the work had only tenuous links with systematic pedagogy. Davy gave prominent acknowledgement to his audience at the Royal Institution, recording that most of the experiments had been conducted in its laboratory, "and all that were fitted for demonstration have been exhibited in the theatre . . . and have been received by the members in a manner which I shall always remember with gratitude."[56] He indeed had reason to be grateful to his audience for validating his experimental findings. It also helped that they were not formal students, and Davy did not have the responsibility of leading them through a comprehensive course of instruction in the discipline—a situation that freed him to take a radically unconventional approach to textbook authorship.

The distinctive feature of the book was that it was closely tied to Davy's own accomplishments as a discoverer. It prominently featured his use of the Royal Institution's voltaic battery to push back the bounds of chemical analysis and isolate new elements. Davy also revived some of his earlier speculations about heat, light, and other imponderable fluids. And he claimed to have subjected all the experiments mentioned in the book to confirmation in his own laboratory. These were departures from the usual practices of textbook authors, who generally believed their duty lay in imposing organization on facts established by others. Some of the book's reviewers realized that Davy had taken advantage of his personal standing to avoid the routine drudgery of authorship. Thomas Thomson, writing in *Annals of Philosophy*, remarked that preparing textbooks normally required a lot of straightforward compilation, rather than original discoveries. That being so, Thomson suggested, "Sir Humphry Davy could hardly execute his task in such a manner as not to diminish his reputation."[57] John Bostock, in the *Monthly Review*, was not the only reader to discern signs of Davy's partisanship toward his own discoveries, a breach of the neutrality normally expected of a textbook author. Bostock concluded that the circumstances "which have attached so much celebrity to the name of the present author, may prove to be causes of imperfections in his works."[58]

In his biography of his brother, John Davy acknowledged that the *Elements of Chemical Philosophy* was "in many respects peculiar." He explained that Humphry as author displayed the same personal characteristics he exhibited in the laboratory, especially the zeal that possessed him when intensely immersed in a series of experiments. John wrote of his brother: "Almost as soon as he began writing, he began printing; no fair copy was made: the MS. was transferred sometimes the same day and hour from his pen to the press." Anticipating the obvious inference, he insisted, "Though rapidly composed, it was not, in fact, hurried."[59] His brother had prepared for the composition by many years' investigation and deliberation, rather as a great artist trained his hand in preparation for the quick execution of a fresco. Others were not so sure that Humphry Davy's genius was really great enough to carry off the project. Bostock discerned in the work "evident traces both of ability and of haste."[60]

The book opened with a historical review of the progress of chemistry, a standard feature of didactic texts of the period.[61] Davy's survey was notable, however, for his idiosyncratic judgments on his predecessors. Black, Cavendish, Priestley, and the Swedish chemist Carl Wilhelm Scheele were said to have been "undoubtedly the greatest chemical discoverers of the eighteenth century."[62] Priestley's characteristics were listed as "ardent zeal and the most unwearied industry."[63] Lavoisier, on the other hand, received barely more than perfunctory mention. "His discoveries were few," Davy commented, "but he reasoned with extraordinary correctness upon the labours of others." There was no denying Lavoisier's importance as a conceptual innovator, and Davy also conceded that, in his use of accurate weight-measurement, "he has entered the true path of experiment."[64] But the French chemist was not a *discoverer*, which was clearly how Davy wished himself to be seen. Perhaps the contrast between British success in discovery and French theoretical speculation was drawn particularly strongly because of the military conflict in which the two nations were engaged at the time. The British prided themselves on originality and freedom, while they saw the French as enchained by Cartesian rationality. But there was also a consistent note of self-projection in many of Davy's comments on the chemists who had preceded him.

After the historical introduction, *Elements of Chemical Philosophy* turned to general thoughts on heat, affinities, electricity, and the role of ethereal substances or imponderables. Davy reiterated his view that heat, or "calorific repulsion," was not a material substance but a motion produced in normal bodies by the influence of "radiant or ethereal matter."[65] Thomson, in his review, objected to this as unwarranted speculation, a frivolous departure from a half century of solid scientific reasoning. Scottish chemists had been upholding the material theory of heat since the 1760s, so it is not surprising that Thomson took issue. Bostock also nailed Davy for his "completely whimsical speculations" on this subject.[66] Less contentious in general was the following treatment of the elements, which were organized into three categories: those that support combustion (oxygen and chlorine), nonmetals capable of undergoing combustion (hydrogen, nitrogen, sulfur, phosphorus, carbon, and boron), and metals (thirty-eight of them, including several that Davy himself had discovered). At the end, there were descriptions of a couple of substances of uncertain status, including the "fluoric principle," which Davy was shortly going to declare another element, similar to chlorine. After that, the volume ended, and no further installments ever appeared. The plan to cover chemical compounds and their uses in the various arts was never fulfilled. Although Davy had told friends that he would bring out another volume the following year, he never did.[67] John Davy clearly felt sufficiently embarrassed about this to offer an explanation, albeit the rather feeble one that his brother was preoccupied with other concerns until illness overtook him in the late 1820s.[68] Unfortunately, Thomson and other reviewers of the first installment had already suggested that the project was hopelessly overambitious and could never be completed.

The reviewers were particularly irked by Davy's attempt to close off further debate about the elemental status of chlorine, or "oxymuriatic acid," as most chemists still preferred to call it. Davy had asserted dogmatically that the traditional view of the gas as a compound was "now universally given up" and had shown how chemical theory could be reconstructed around the assumption that it was an element.[69] The reviewers reminded him forcefully that this was not an accurate representation of the still-ongoing debate. Bostock wrote that Davy's "opinion" was "very ingenious, and perhaps correct, but it is certainly not yet demonstrated." Readers would know that John Murray was still defending the contrary view and would therefore regret that Davy "has permitted himself to speak in so confident a manner on this subject."[70] Even an unnamed reviewer in the conservative *Quarterly Review* declared that the tone of the author's remarks on chlorine was "somewhat more decisive than the present state of the investigation altogether authorizes."[71] It seemed to these commentators that Davy was finding it difficult to set aside his advocacy of his own discoveries, as he should have done to discharge the duty of a textbook author.

Underlying the hostile responses of the reviewers were their ideas—disparate from Davy's—about what was expected of the writer of a textbook. Thomson, Bostock, and others in the British chemical community associated authorship of a textbook with adoption of a neutral stance on controversial issues. Nicholson had provided an example of this, with Priestley in the background as a more distant inspiration for this kind of writing. One way neutrality could be signaled was by adopting a historical narrative, describing empirical facts in the order in which they had been ascertained, without making judgments on their relative significance or theoretical interpretation. This was the style of writing Priestley had adopted and which Thomson used in his own *System of Chemistry* (1802). Other textbook authors of the time, including Samuel Parkes, William Henry, and Murray himself, declared that they were leaving to readers the judgment of controversial issues. For this reason, they were cautious about new terminology, which might bias students toward a particular theoretical view. Murray referred to "the example so evident of the evils attending a theoretical nomenclature."[72] He was thinking, no doubt, of the terminology proposed by Lavoisier and his colleagues in 1787, which had been so fiercely resisted by British chemists. Davy, on the other hand, was willing to stamp his personal imprint firmly on the parts of the discipline he treated. He acknowledged the need to maintain neutrality in terminology, but he claimed he had the support of his most distinguished peers for using the name "chlorine," and he went on to suggest further new names for its compounds. The innovations aroused objections from other chemists, which Davy may have planned to address in later editions of the *Elements of Chemical Philosophy*.[73] But he believed he was entitled to modify nomenclature in line with what his experiments had established. To the discoverer belonged the prerogative of naming his discoveries.

Davy's self-presentation as a textbook author was thus closely linked with the public persona he had fashioned as a discoverer. But there was also another component of his authorial identity, which emerged in this book in particular. The word "philosophy" occurred in the title, and Davy clearly intended to appear in the text as a *philosopher*.[74] This intention came through most clearly in its final section, where he stepped back from his assertions about the impossibility of decomposing chlorine to admit, "We know nothing of the true elements belonging to nature."[75] To say that chlorine and the other simple substances could not be decomposed by any currently known methods did not mean that they could never be analyzed by some method yet to be discovered. This admission opened the door to speculation as to what the ultimate constituents of all matter might be. Perhaps, he suggested, hydrogen "approaches nearest to what the elements may be supposed to be," and everything else might be compounded from it.[76] Such a thought evoked the metaphysical doctrines of "the ablest philosophers" of ancient times, who had taught the fundamental unity of matter. It was a "sublime chemical speculation, [also] sanctioned by the authority of Hooke, Newton, and Boscovich."[77] In fact, Davy had been indulging in these sublime speculations for a while. In March 1809, Dinwiddie heard him discuss the possibility that all substances might be reduced to different arrangements of just one kind of matter, a prospect he labeled "the dream of Newton."[78] In the introduction to the *Elements of Chemical Philosophy*, he explained that reducing the diversity of chemical objects to an underlying uniformity would reveal the simplicity of the divine design. The sublime chemical philosophy thus buttressed the claims of natural theology to reveal God's wisdom in the creation.

There were tensions between the two ways Davy presented himself in this work. On the one hand, he demanded credit for his discoveries, especially the products of analysis yielded by his pioneering application of the voltaic pile. On the other hand, however, he wished to be seen as a philosopher, an image that allowed him to speculate about the ultimate components of matter, beyond the limits of any available analytic methods. The contrasting roles reflected a transition at the point in Davy's life when the *Elements of Chemical Philosophy* appeared, coinciding as it did with his marriage, knighthood, and retirement from the Royal Institution. As we shall see in the next chapter, the status of an independent gentleman of science, which he subsequently assumed, went along with his increasingly frequent self-representation as a philosopher.

The changes in Davy's circumstances after 1812 no doubt played a part in his abandonment of the textbook project. Even the portion that did appear never won widespread acclaim, and no second edition was ever called for, largely because of a disjunction between Davy's presentation of the subject and the pedagogical needs of the chemical community at large. Opportunities for teaching and learning chemistry were flourishing as never before, particularly in London. Courses of instruction were offered in medical schools, in hospitals, in private premises, and in the numerous institutions that sprang up at the time, including mechanics' institutes and learned societies. After the passage of the Apothecaries Act in 1815, medical students were required to study chemistry. But those who were involved in teaching the subject found Davy's book poorly suited to their needs. The textbooks of Nicholson, Thomson, Henry, and Murray retained their position in the market, unchallenged by Davy's innovative alternative.

This is not to say that Davy was not highly regarded for his scientific accomplishments. What the Scottish writer Andrew Ure called his "predominating genius" was an unavoidable figure on the scene.[79] Even authors who differed from him on points of theory acknowledged his greatness as a discoverer. And his reputation among the general public was unassailable, buttressed by popular expositions of his work in dialogue form, such as Jane Marcet's *Conversations on Chemistry* (1806) and Jeremiah Joyce's *Dialogues in Chemistry* (1807). But, precisely by celebrating his genius for discovery, such works contributed to the sense that he was extraordinary and impossible to emulate. Even Faraday, who became the nearest thing to a student Davy ever had, ended up working in other areas of science and building an entirely different public profile. Davy's talents lay more in constructing his own persona than in rallying a community around himself. His abilities and the resources he commanded were so obviously singular that he provided no model for younger chemists to aspire to. His influence on how chemistry was taught was in the end limited and largely mediated by others. Davy's story shows just how difficult it can be to capitalize on individual genius to reform the structure of a scientific discipline.

5. The Philosopher

If I were engaged in any high undertaking or design, fraught with
extensive utility to my fellow-creatures, then could I live to fulfil it.
MARY SHELLEY, *Frankenstein*

The characters that Humphry Davy took on through the course of his
life were the result of how he was seen by others and of his own self-
fashioning. His public profile both reflected and reinforced a process
of introspective self-formation. He thought, for example, that he might
be becoming an enthusiast as he realized the implications of his experiments with nitrous oxide. Later he seems to have played up to some of
the expectations of that role. As a young man, he nurtured the ambition of being a genius, and in his subsequent public performances he
found ways to embody the type. He would never have described himself as a dandy, though the label seemed apt to many of his contemporaries. On the other hand, he insisted on his identity as a discoverer,
cultivating the authority that derived from having found several new
chemical elements.

Davy often accepted for himself the straightforward label "chemist," but he also sought recognition as a "philosopher." To call himself a philosopher meant several things. It implied a grasp of the more
profound or sublime questions of chemistry, such as whether all substances were ultimately composed of the same kind of matter. When
he lectured about such issues, he did so under the heading of "philo-

sophical chemistry" or "chemical philosophy," the latter phrase also appearing in the title of his abortive textbook project. These philosophical lectures were distinguished from the concurrent series on experimental and practical chemistry, which did not share the same lofty agenda. When discussing the finer points of matter theory, Davy presented himself as a philosopher, as James Dinwiddie witnessed. Davy declaimed on the cosmological and geologic roles of heat, he quoted passages from the Roman poet Lucretius, and he pondered what he called the "dream of Newton" concerning the underlying unity of all matter.[1] To be a philosopher was to nurture these profound metaphysical interests.

The philosophical chemist also claimed to have theoretical knowledge that was relevant to many practical activities. During the decades of the Enlightenment, chemists who called themselves philosophers had asserted their authority over many fields of technical skill: metallurgy, mineralogy, brewing, agriculture, bleaching, pharmacy, and the distilling of spirits, for example. Eighteenth-century authors and teachers, including Peter Shaw in England and William Cullen in Scotland, insisted on the superiority of theoretical knowledge over that of the laboring artisan.[2] On the basis of his knowledge of fundamental principles, the philosophical chemist asserted his right to direct the activities of those whose knowledge was purely practical or empirical. To be a philosopher meant laying claim to this authority over those who pursued the chemical arts.

Davy built on this eighteenth-century conception of the relations between chemical theory and practice. But his authority as a philosopher in these fields could never be taken for granted, as we might tend to assume if we identify the chemical arts too readily with "applied science" or "technology." It was never easy for Davy to translate his theoretical knowledge into practical achievements, in part because the applicability of chemical theory was not universally accepted in the fields in which he worked. Rather, he had to build his authority gradually through the experience of grappling with practical problems. He did so repeatedly throughout his career, devoting himself to a series of investigations of such topics as the tanning of leather, agricultural chemistry, gunpowder manufacture, the design of the miners' safety lamp, and the protection of ships' hulls from corrosion. In each domain, Davy's status as a philosopher was both a weapon to wield and a prize for which to contend. When he succeeded in improving the prevailing practices in the field in question, Davy consolidated the authority of theory over practice and simultaneously burnished his own credentials as a philosopher. But his success was often contested, and the disputes were sometimes prolonged. In some cases, Davy secured credit for an innovation only after lengthy controversy, and the relevance of theory to practice consequently remained in doubt for some time. Simplistic or anachronistic assumptions about the dependence of technology on science cannot guide us through these complex debates because they prejudge the outcome.[3] In fact, as we shall see, Davy had to struggle in each instance to establish that the theoretical knowledge of the philosopher could have practical benefits.

To be a philosopher meant something else as well. Since classical antiquity, it had implied a degree of detachment from the cares of the everyday world. This made for additional complications and something of a paradox, as Davy pursued his inquiries into the chemical arts. As Francis Bacon put it in the seventeenth century, the philosopher's difficulty was to balance the demands of *negotium* (involvement in the affairs of the world) with *otium* (the seclusion required for contemplation).[4] A philosopher was supposed to use his wisdom to benefit humanity in general but at the same time to be detached from ambitions of personal gain. He was not supposed to be motivated by self-interest but rather by a general sense of benevolence. Davy seems to have become more insistent on this dimension of his philosophical identity after his marriage and retirement from paid employment at the Royal Institution in 1812. In the projects in which he was involved during the 1810s and 1820s, he jealously guarded his philosophical persona, vigorously disavowing motives of personal monetary gain and insisting on the purely intellectual and disinterested character of his contributions.

There was an obvious social dimension to this aspect of Davy's self-fashioning. As a philosopher, he represented himself as categorically distinct from an artisan, a businessman, or a tradesman. In these years, Davy laid unambiguous claim to the identity of a gentleman and an aristocrat, with a knighthood followed by award of a baronetcy. As he modeled himself on the aristocratic grandees with whom he associated, any taint of mercenary or commercial ambitions would have been anathema to the character he was forging. At this stage of his life and after his death, Davy was often accused of having toadied to the aristocrats with whom he socialized. No doubt, there was some basis for that charge. But assuming the character of a philosopher allowed him to portray his social metamorphosis in a more respectable light. Not merely a social climber, he represented himself as a pure intellectual, a man of wisdom, whose profound theoretical understanding was placed selflessly at the service of humankind.

Davy was not the only individual to experience the contradictions of the philosophical identity at this time. As David P. Miller and Christine MacLeod have shown, becoming known as a philosopher in connection with scientific innovations was a complex matter. It meant staking out a position in negotiations over social standing and material advantage, and it could have problematic consequences for the individual who took on the role. The Scottish chemist and industrialist James Watt asserted his identity as a philosopher to insist on his intellectual priority in relation to certain inventions and to claim a more respectable place in society.[5] The prestige of the label did not long survive in the nineteenth century, at least not in connection with technical achievements, since the new identity of "engineer" soon succeeded it. But being known as a philosopher was a valuable social strategy in the period when Watt and Davy made their careers. It is telling that Davy endorsed Watt's choice of moniker, putting his own spin on what constituted the Scottish chemist's claim to philosophical status. In 1821, he wrote to Watt's son James that his recently deceased father was a "truly illustrious philosopher."[6] The same term featured prominently in the eulogy Davy delivered in June 1824, when a monument to Watt was erected in London. On that occasion, he stressed that the Scottish industrialist was no mere "practical mechanic" but a "natural philosopher and a chemist." He instanced Watt's renowned improvements in the design of steam engines, which were said to be rooted in a profound knowledge of the science of heat as taught by the famous Scottish chemist Joseph Black.[7]

Davy recognized Watt as a kindred philosophical chemist, but his own background and accomplishments were different. He was not able to attach himself so directly to the traditions of pedagogy and practical improvement that originated in the Scottish Enlightenment. Davy operated in a different context and consequently faced different challenges in the course of his involvement with the chemical arts. Calling himself a philosopher had strategic value to him as he tried to negotiate these challenges, but it turned out to yield rather mixed success.

The word "philosopher" captured some of Davy's most intimate and profound intellectual ambitions, from early in his life to his last days. Gregory Watt, the friend of his youth, addressed him in their bantering correspondence as "Amiable Philosopher" and "My dear Philosopher."[8] In turn, Davy assigned the label to some of his early mentors, such as Samuel Taylor Coleridge and the Cornish man of science Davies Giddy, making it clear that he aspired to the same designation. When he shared his speculations on light and heat with Giddy in February 1799, he wrote that he did so out of deference to Giddy's standing as a philosopher.[9] In the following year he labeled Coleridge a "poet philosopher," a term he was also applying privately to himself at the time.[10] Years later, as he was preparing his last book, the series of dialogues that became *Consolations in Travel* (1830), he deliberated at length about the characters to be included in the narrative, one of whom was originally to be named simply "Philosopher."[11] When it appeared, after Davy's death, the book was subtitled *The Last Thoughts of a Philosopher*, and it contained a chapter called "The Chemical Philosopher." Understandably, then, John Davy judged that this was the most important of the many personae his brother had adopted in the course of his life. John concluded that it was this identity that tied together all of his brother's many roles: "As such he began his career[,] as such he terminated it & as such I trust his name will descend to posterity."[12]

To assume the title of philosopher was to lodge a claim to profound intellectual understanding and perhaps to a kind of intuitive insight into nature. It also connoted a certain moral standing, traditionally to be achieved by a regime of self-discipline and rigorous self-control. Humphry Davy knew about the classical traditions of philosophical ethics and was sometimes given cause to measure his own behavior against the strenuous expectations they set. The nitrous oxide investigations, for example, raised the question of whether the appropriate moral standards had been upheld while Davy and his friends were indulging in drafts of the "pleasure-producing air." In a letter to his Penzance patron Dr. John Tonkin, in which he described these experiments, Davy was careful to note that they had been endorsed by "the most respectable of the English philosophers."[13] His early years in London posed another kind of ethical challenge, with the lure of dissipation and the distractions of fame. Coleridge was not the only one of his friends to worry whether his philosophical virtue would be compromised by these temptations. Davy himself reflected on the perils in a draft of another letter to Tonkin. He acknowledged, "Temptations speak every where to men in great cities which are the abodes of vice & of indolence." As a protection against these hazards, he promised to keep his mind active and focused on "pursuits useful to mankind . . . which promise to me at some future time the honorable meed of the applause of enlightened men."[14]

The formal and slightly stilted language here captured both the dangers Davy faced in his position at the Royal Institution and the means by which he hoped to defeat them. His new situation promised opportunities to act benevolently toward humanity as a whole. If he could make the most of them, he could derive moral advantage from the resources of the capital and offset the perils of London life. In the same notebook where he drafted the letter to Tonkin, he also jotted some remarks on the progress of science and social improvement. Some of these comments found their way into the "Discourse Introductory to a Course of Lectures on Chemistry," delivered in 1802. But there was a sharper edge to some of Davy's private remarks, which was blunted when the speech was given in public. In both documents he referred to "the rich and privileged orders" of society as the guardians of refinement, civilization, and science itself.[15] In his notes, he coupled this with a sharply worded criticism of these elite groups, where, as he put it, "dissipation & the seeking after unnecessary wants has too much occupied the mind." The separate classes within society had, he suggested, "been too much insulated in small circles of self interest." What was needed was to connect them "in a state [of] society in which the character of the philosopher was united to that of the artist."[16] When the discourse was delivered in the Royal Institution's lecture theater, Davy suppressed the criticism in favor of a more complimentary image of the upper classes. Now, he said, they were "giving up many of their unnecessary enjoyments in consequence of the desire to be useful." With the establishment of the new institution, he suggested, the leaders of society were beginning to shoulder their responsibilities as "friends and protectors of the labouring part of the community."[17] One could therefore look forward to a productive relationship between philosopher and artisan, with the former delivering the scientific principles by which the latter could be guided toward improvement.

Davy was making the case for the Royal Institution as a site where philosophical virtue could be exercised for the benefit of humanity at large. Of course, this flattering portrayal gratified the self-image of the institution's proprietors. Count Rumford, Sir Joseph Banks, and the others involved in the foundation saw its philanthropic activities as social duties, and also as bulwarks of the aristocratic hegemony against subversion.[18] After a decade of political turmoil and social unrest, Davy was no doubt right to think they would not wish to hear criticisms of the upper classes. Rather, it was politic to laud the sense of social obligation that the new establishment expressed. The Royal Institution was a place where scientific knowledge was to be allied with political power. But it was not entirely hypocritical to describe the situation in terms of moral behavior. Davy was figuring out how to negotiate his own place within the new institution, how to serve his own interests while also developing his identity as a philosopher. To do so, he needed to provide his patrons with the scientific knowledge on which improvements in the practical arts could be built. And he also needed to align his sense of philosophical virtue with the prevailing aristocratic ethos of social duty.

His first substantial assignment to explore the possibilities of scientific improvement came in June 1801, when the managers of the Royal Institution instructed him to prepare a series of lectures on the chemical principles of tanning and gave him leave for the summer months to inform himself about it.[19] The topic was of pressing economic importance because of the shortage of oak bark, the traditional source of the tannin used to impregnate animal hides to make leather. A systematic inquiry into the underlying chemistry of the process, and an exploration of alternative possible sources of tannin, seemed to be required. When he mentioned the subject in the "Discourse Introductory," Davy was inclined to lay the blame for lack of progress on intractable laborers. The methods of tanning had, he claimed, already been "reduced to scientific principles." The obstacles that stood in the way of improvement arose from "the difficulty occurring in inducing workmen to form new habits." It was a case of "the common prejudice against novelties" obstructing the efforts of "chemical philosophers."[20] The following year, however, when his report on the subject appeared in the *Philosophical Transactions*, he seemed to have become more receptive to the value of artisanal knowledge. He announced a significant discovery—a new tanning principle extracted from the bark of an Indian mimosa tree—but he also acknowledged there was some truth in "the vulgar opinion of workmen" about the time necessary for the operation to occur.[21] It would not be possible to speed up the impregnation process significantly. In this respect, at least, the practical knowledge of laborers could not be improved on by theoretical inquiry.

Davy's investigation of tanning was, in certain respects, a chastening experience, even though he received the Royal Society's Copley Medal in recognition of his efforts. He seems to have retreated somewhat from his initial confidence that chemical investigation could introduce significant improvements. The lesson was partly learned through his association with skilled tanners whom he respected, especially Samuel Purkis and Thomas Poole. Purkis, owner of a factory at Brentford in Middlesex, supplied important information to Davy and was acknowledged in the 1803 publication. He became a friend with whom Davy kept in contact for years. Poole's friendship was even more long-lived, from the time of Davy's youth to the end of his life. Poole owned a tannery and farm at Nether Stowey in Somerset, where Coleridge was for a while a frequent visitor. The two friends joined Davy in Bristol to partake of nitrous oxide, and they later traveled to London together to see him lecture. In the late 1820s, after Davy's health began to fail, he made two visits to Nether Stowey in attempts to recuperate.[22] When *Consolations in Travel* made its posthumous appearance, it bore a dedication to Poole, "in remembrance of thirty years of continued and faithful friendship."[23] Poole, more than anyone, represented to Davy the figure of the philosophical artisan, an individual who combined in a single person scientific wisdom, devotion to humanity, and skill in the practical arts.

Poole was also said to have been one of Davy's mentors in agriculture; he took a deep interest in his young friend's inquiries in this domain.[24] But Davy's agricultural chemistry also brought him into contact with much higher-ranking individuals, who presented him with more prominent examples of philosophical virtue. He first gave lectures on the subject in 1802, appointed by the Board of Agriculture, many of whose aristocratic members were also involved in the foundation of the Royal Institution. In the following years, leading up to the publication of his *Elements of Agricultural Chemistry* in 1813, he often visited the estates of the landowners who were sponsoring his work, including that of John Russell, the sixth Duke of Bedford, at Woburn, and that of Sir Thomas Bernard at Roehampton. He investigated the effects of various manures and fertilizers on experimental plots at both sites. In a large-scale painting by George Garrard, he is shown attending a sheepshearing at Woburn in 1804 and conversing with Banks, with Sir John Sinclair (president of the Board of Agriculture), and with the agricultural improvers Thomas Coke and Arthur Young.[25]

Bedford and Coke (a member of Parliament with substantial land-holdings in Norfolk) were prominent Whig grandees who advocated agricultural improvement within the established social hierarchy. They hoped to serve the British nation by increasing food production, thereby saving it from revolutionary upheaval or the dire fate of starvation notoriously foreseen by Thomas Malthus. The ambition was widely shared among elite leaders of the Whig Party. Improvements in estate management were seen as consistent with a liberal and enlightened outlook, a component of the Whig ideal of gentlemanly identity. They could also be connected to the "Georgic tradition" of rural literature, which descended from the revival of interest in Virgil's pastoral poetry in the middle of the eighteenth century.[26] The prevalence of this outlook among Davy's patrons allowed him to make the pitch for the agricultural utility of chemical knowledge, but it also tended to limit the role ascribed to technical expertise within the enterprise.

Davy's self-modeling as a philosopher was of key importance. Arguing for the relevance of philosophical chemistry to agriculture, he drew upon the precedent of Scottish chemists from earlier decades, especially the Edinburgh professor Cullen. Davy's language in the "Discourse Introductory" resonated with Cullen's. He declared that the activities of the agriculturalist would be "profitable and useful to society, in proportion as he is more of a chemical philosopher."[27] Even while performing soil analyses as a condition of his employment at the Royal Institution, Davy was trying to persuade landowners to take up chemistry themselves. He wanted to convince them that chemical knowledge could be of philosophical import and thus consistent with their self-conception as liberal and enlightened gentlemen. This attempt at persuasion was done, in part, by articulating the landowners' own paternalistic model of improvement. Davy told the audience at his lectures on agricultural chemistry that

it is from the higher classes of the community, from the proprietors
of land; those who are fitted by their education to form enlightened
plans, and by their fortunes to carry such plans into execution; it
is from these that the principles of improvement must flow to the
labouring classes of the community.[28]

Paying tribute to his audience's ability to grasp chemical theory,
Davy set out his ideas on plant nutrition and respiration. He drew his
audience into his long-standing speculations on the role of light in the
processes of life. On a more mundane level, he discussed the efficacy
of crop rotation and the uses of various manures, and he promoted
a specific protocol for chemical analysis of soils. This last was a sim-
plification of the routine advocated by earlier writers, including the
Irish chemist Richard Kirwan and the Scottish agricultural improver
Archibald Cochrane (Lord Dundonald).[29] Davy claimed that the skills
and equipment required to determine the chemical composition of
soils could be quite modest, within the capabilities of any gentleman
farmer. The procedure would be carried out by evaporating a sample
taken from the field, conducting various titrations and precipitations,
and carefully weighing the products. The analyst would require a small
collection of apparatus, including a sensitive balance, an Argand lamp,
some glassware, various vessels, and selected reagents. But the whole
thing could be encompassed within a small closet. There was no need
for a large laboratory or for specialized training. Davy was insistent
that the methods could be carried out by a gentleman improver himself, without calling on any additional expertise.[30]

When the *Elements of Agricultural Chemistry* appeared, some of
the reviewers wondered whether the author had unrealistic expectations of the chemical skills farmers could command. The *Edinburgh
Review* pointed out that the analysis of vegetable matter that Davy had
described was much too demanding for anyone but an expert chemist to perform. This reviewer also noted that the book was priced too
high for "the practical husbandman" to afford.[31] Nonetheless, the publication was a great success. Later editions appeared in cheaper for-

mats and sold well, including in Ireland and the United States. The book held its place in the market until it was displaced by an English translation of Justus von Liebig's work on agricultural chemistry in the 1840s. Up to that point, at least six editions of Davy's book appeared in the United States, with many synopses and adaptations aiming to capitalize on its reputation. It was widely adopted by agricultural improvement societies in many states. Davy's model of the gentleman farmer who was open to enlightened scientific improvement chimed with the Jeffersonian ideal of a republic of virtuous and liberal-minded landowners.[32] The *Edinburgh Review* had already noted the American resonances of this model, remarking that George Washington himself ("the great founder of American liberty") had exemplified the pattern of a part-time military and political leader whose deepest commitment was to farming the land.[33] Americans who viewed their first president as a paragon of republican virtue were well disposed to adopt Davy's suggestions for improving agricultural productivity.

Davy had hit on a successful formula. Deferring to the enlightened paternalism of his elite patrons, he had incidentally struck a chord with an international readership of improving farmers. His articulation of the image of the chemical philosopher was one reason for this success. Davy was calling for improvements in the rural arts, but he was not explicitly making a case for the indispensability of specialist expertise. He was not proposing that every agriculturalist should hire a trained chemist to work for him. In contrast to the case of electrochemistry, where he aimed to centralize expertise and resources, Davy wanted agricultural chemistry to be practiced by enlightened landowners themselves. By acquiring a knowledge of chemistry and a modest collection of apparatus, the gentleman improver could improve agrarian productivity and, at the same time, exercise the virtues praised in the Georgic tradition. This outlook echoed the liberal self-image of the Whig grandees who originally sponsored Davy's work, and it also turned out to have a much wider appeal.

As Davy continued to be involved with schemes for improvements in the chemical arts, his model of philosophical virtue was put to the test. Time and again, his self-image as a philosopher who benevolently passed down scientific principles to those who put them into practice was called into question. Each new problem required a new inquiry into the relevant scientific principles and how they should be applied. The distribution of authority between the man of science and the practitioners of the arts had to be renegotiated each time. And, on each occasion, Davy forcefully asserted his independence from motives of personal gain. The issue became a more pressing one after his marriage brought him wealth and allowed him to retire from salaried employment. From this point on, he clearly believed he had earned the badge of gentlemanly status with its associated economic independence. In all his scientific projects, he became particularly eager to establish that his motives were higher and purer than monetary ones.

The issues were brought to the fore during his involvement in the gunpowder manufacturing enterprise of his friend John George Children.[34] Children, the son of a wealthy banker, lived at Ferox Hall in Tonbridge, Kent. He was an accomplished chemist, with a long-standing interest in pyrotechnics, who built a large voltaic battery for electrochemical experiments. He and Davy also shared an interest in hunting and angling, and the gunpowder project seems to have originated from their common passion for shooting. The two friends often discussed the design of guns and the effectiveness of various powder formulations. In 1811, Davy began to investigate the proportions of the components in gunpowder, convinced that the traditional mixture of saltpeter, sulfur, and charcoal could be improved if one understood the law of combining proportions and the chemical reactions involved in an explosion. In July 1812, he wrote to Children that "A *perfect* gunpowder" would be composed of the following proportions: 191 of saltpeter (or potassium nitrate), 28.5 of charcoal, and 30 of sulfur. In the same letter, he noted that he had returned to Children's solicitor a signed agreement formalizing his participation in the project to manufacture powder according to this formula.[35]

June Fullmer, who published a study of the whole affair, concluded that Davy at this stage was hoping to profit from the enterprise but that he quickly changed his mind. She noted that he wrote around the same

time to the publisher of his *Elements of Chemical Philosophy*, request-
ing the deletion of a passage on the lead-chamber process for making
sulfuric acid. Apparently, he believed he had discovered how to improve that process, using a similar approach to the one applied to the
gunpowder question, and he contemplated taking out a patent on it.[36]
In October 1812, Davy was still writing to Children of his "ardour" to
manufacture gunpowder, and he visited the Tonbridge works later that
month.[37] His attitude seems to have changed rather suddenly after he
saw one of the advertising flyers Children had prepared to promote
the product. In June 1813, he wrote that he was concerned with the
wording on them, which he thought could damage his reputation.[38]
The following month, after another trip to Tonbridge, he became even
more anxious. He dispatched a series of increasingly urgent letters to
his friend, demanding that the flyers be altered and his role in the partnership canceled. On 19 July 1813, he wrote to Children from Rokeby
Park, the home of the antiquarian and member of Parliament John
Morritt in County Durham:

—I have been much disturbed & vexed by enquiries respecting the
price of *my* gunpowder which from the labels I find is supposed to be
sold by me. — These labels must be altered so as to put in a clear point
my relations to the manufacture & it must be understood by the public that I have given my gratuitous assistance & advice only . . .

—In the labels on the windows it should not be *under my directions*
for this implies that I am a superintendent in the manufactory. But
"Ramshurst extra superfine or other Gunpowder manufactured by
Mess. B.C. & Co. In the composition of this powder the proprietors
have or are been assisted by the advice & exp[ts] of Sir H Davy."[39]

The crucial thing that "disturbed & vexed" Davy was the suggestion
that customers would think he had a commercial interest in the product, that it was being sold by him or made under his supervision. He
insisted that it be stated that his advice to the manufacturers had been

given freely, without payment or financial interest in return. The matter continued to weigh heavily on his mind. A few days later, after he had reached Edinburgh, he wrote again that he had been "in extreme harass and anxiety" about the wording on the labels of the canisters in which the powder would be sold. All the friends he had consulted had advised him that it was "absolutely necessary for my reputation" that it be stated that he had given his advice gratis. He concluded: "I have resolved to make no profit of any thing connected with Science—I devote my life to the public in future & I must have it clearly understood that I have no views of profit in any thing I do."[40] The following day, writing from Scone in Perthshire, he again berated Children to alter the wording on the labels and advertisements: "till the alterations are made . . . I shall be miserable."[41]

The frantic, pleading tone of this correspondence shows Davy in an unfamiliar light. This was not the dazzling genius in full command of his instruments of discovery, nor was it the glamorous young man who captivated his audiences at the Royal Institution. Instead, the writer seems wracked by a sense of insecurity, even uncertainty as to his own identity. Davy obviously felt that his reputation was at stake and his standing in society at risk. It was as if his personal character would be fundamentally impugned by the suggestion that he was to derive any profit from the project. Fullmer suggests that Davy's recent marriage was a significant factor in heightening his status anxieties at this time, that he was sensitive to the gossip that suggested he had married for money and particularly worried about protecting his reputation for probity.[42] There is probably something in this suggestion, though it should be noted that Davy had been married more than a year before he raised his concerns about the advertising of the gunpowder. Indeed, he was already married when he entered into the partnership with Children and his collaborators, a partnership he subsequently demanded be dissolved and concealed from the public.

Children and the others involved in the enterprise found Davy's change of heart capricious and incomprehensible. They ascribed his anxieties to an excessive sense of his own importance. Another partner in the scheme (a Mr. Burton) wrote to Children, "Public attention . . . has so much inflated this Young Man, that . . . the final result will be to tarnish the lustre of his outset." He was incensed that Davy had "lent his name to a concern, and committed the parties engaged in that concern, not only the most deeply in their fortunes, but in their Character with the Public—and then upon grounds as frivolous as they are unjust, . . . deserted it."[43] To Davy himself, Burton wrote in more conciliatory terms—that "it never was our intention to found the manufactory on assertions that wo[ul]d affect your Credit, scientifically or otherwise"—but the sense of bafflement and affront was clear.[44] Also apparent was the suggestion that the partners' fortunes and financial credit were at risk and perhaps rendered more vulnerable by Davy's behavior. Fullmer proposes that Davy might have had an inkling that the Children family banking enterprise was insecure, and indeed, it collapsed two years later.[45] It is possible that a foreshadowing of this danger provided another incentive for him to withdraw from the project.

Whatever the immediate circumstances of Davy's about-face on the gunpowder enterprise, his action was consistent with his lifelong project of self-identification as a philosopher. The reputation he was concerned to protect was bound up with a pose of lofty detachment from worldly cares and ambitions. He made this concern over his reputation especially clear in another letter to Children, on 26 July 1813, in which he appealed for a discharge from the partnership agreement into which he had entered. He declared that he would go abroad in the autumn on "a very long journey" and was anxious to settle all his affairs before departure. He went on:

> I depend upon you to set my mind entirely at rest, I have had great anxiety & much mental pain from the idea that my philosophical repose & usefulness might be destroyed by the cares of business or the trouble of litigation. . . .
>
> Science & philosophical quiet are my great objects & these are the motives which the world must attribute to me in withdrawing from all commercial speculation.—Were I to do otherwise I must be a bad reasoner & unworthy of the name of philosopher.[46]

There was a degree of disingenuousness in Davy's claim that he was seeking "philosophical repose." He was certainly not abandoning the ambition of "usefulness," or finding ways to apply his scientific knowledge to practical ends. But, with his newly acquired knightly status, it was vital that "the world" attribute the right motives to him in doing so. Association with commercial speculation or any suggestion of involvement in business was antithetical to the pose he was trying to strike. At this juncture, the persona of a philosopher helped him secure his independent standing. He claimed to be looking for "philosophical quiet" and seclusion. He worried that he would be thought unworthy of the name of a philosopher if he were known to have a financial stake in one of the projects he was helping advance.

The same self-image and the same anxiety about protecting it were evident on the best-known occasion when Davy applied scientific knowledge to practical ends—namely, his invention of the miners' safety lamp. The invention, announced in December 1815, became the most celebrated of Davy's discoveries. It provided miners with a source of illumination that was safe to use in the presence of flammable gases underground. Davy declined to take out a patent on the lamp but insisted on being accorded credit for having invented it and making it available to humanity at large. Portraits showed him with the lamp as his personal attribute, a symbol of his disinterested concern for human welfare. This altruistic image was a further boost to his credentials as a philosopher. He was being depicted as one who turned his knowledge to practical ends without seeking personal material gain. Such an accomplishment was not, however, inherent in the discovery itself. Rather, it was the outcome of a series of struggles in which Davy fought to define and secure credit for his invention. He had to work to gain recognition of the device as the product of his own scientific genius, while at the same time disclaiming any proprietary rights over it. Only in this way would the lamp appear as an appropriate attribute of a true philosopher, which was how he wished to be seen.

Davy began his work on the safety lamp in the summer of 1815, after returning from the Continental tour on which he had embarked nearly two years earlier. A group of mine owners from the Northeast of England approached him to apply his expertise to the problem of designing a lamp that would not ignite firedamp (predominantly methane), the combustible gas often found in coal mines. In the design he eventually made public, the lamp's flame was surrounded with a fine wire mesh, which allowed the light to shine out but prevented ignition of any surrounding gas. The path to the final design was by no means straightforward, and it became a subject of contention as Davy's credit for the original conception was disputed. Allegations of plagiarism were leveled against him by rival claimants to the invention, and Davy responded in kind with accusations that his own design had been stolen by his competitors. As the dispute became more intense, stipulations as to specific features of the lamp and when they had been arrived at were retrospectively adjusted, as the contending parties produced their own self-serving narratives.

Frank James has given a careful reconstruction of the whole affair, using surviving manuscript and printed evidence, along with the artifacts held in various collections. He has pieced together the path taken by Davy in his work on the lamp from October 1815, noting that the record is at variance with the "ambiguous, if not misleading, statements" the chemist later made.[47] In an intensive period of work during the last three months of the year, Davy used the laboratory resources of the Royal Institution and drew on the assistance of Michael Faraday to prepare a series of designs. Several of these designs were executed in practice and others merely sketched. One version of the device was announced in a paper read at a meeting of the Royal Society on 9 November 1815, but that manuscript has not survived. The evidence indicates that Davy fundamentally changed his conception of the design after this presentation, abandoning the idea of supplying air through narrow tubes and instead placing a fine wire mesh or sieve around the flame. This was the principle used in the mesh cylinder design that was published, first in the *Philosophical Magazine* in late December 1815 and then in the Royal Society's *Philosophical Transactions* in January 1816. It was soon put to the test in mines belonging to Davy's patrons in Northumberland and County Durham.

By this point, claims were already being lodged on behalf of alternative designs for a miners' lamp, with the suggestion that Davy had failed to acknowledge his debt to these precursors. George Stephenson, an engineer at the Killingworth colliery near Newcastle-on-Tyne, had been trying his own prototype safety lamp during the months of October and November 1815, when Davy was visiting mines in the area. Another design, by the Irish physician William Clanny, involved a more cumbersome arrangement for supplying air to the flame through bellows. It had been in use in Northumberland mines for a couple of years by that point. As Davy asserted his originality and priority over these competitors, it became crucial to insist on the scientific or philosophical principles underlying his own device. Although he actually worked his way toward his final design by an empirical process of trial and error, Davy repeatedly maintained that he had been guided by scientific knowledge and an elevated philosophical viewpoint. In a set of notes in which he rehearsed his case, he described Stephenson as "a mechanic" and condescendingly referred to the engineer's "experiments if his can be so called." Stephenson's claim was being upheld by local worthies whom Davy denounced as ignorant, while he called his own supporters "the most distinguished scientific men in England."[48] He could draw support from a report on Stephenson's device that appeared alongside his own paper in the *Philosophical Magazine* in December 1815. There, an anonymous author (actually the Newcastle botanist Nathaniel Winch) wrote that Davy's invention "flowed from science judiciously applied," whereas Stephenson, "though an excellent mechanic and acute man, . . . is unacquainted with the science of chemistry."[49]

As the dispute unfolded, charges of plagiarism were made on both sides. Stephenson claimed that Davy's wire mesh design was just "a variation in construction" on his own.[50] Davy insisted, somewhat implausibly, that he had no knowledge of Stephenson's device until after his own design was published. Each recruited supporters from mine owners in the region, and the controversy assumed a party-political coloration. Davy's case was upheld by a group led by the Whig magnate

Lord Lambton, later Earl of Durham, who as young John Lambton had lodged and studied with Thomas Beddoes in Bristol.[51] Many of the local Tories supported Stephenson; and when Davy was awarded a service of plate by Lambton and other mine owners in gratitude for his invention, they raised a reward for their own man too. Davy's response to the recognition accorded his competitors was less than generous. Addressing Lambton and his friends when they convened on the celebratory occasion, Davy mentioned "some attempts that had been made to pirate his invention, and some mean insinuations against its originality."[52] When the Society of Arts in London awarded a medal to Clanny, Davy referred to them as "a Committee of Tradesmen" without standing among "Men of real Science."[53] After the announcement of the award to Stephenson, he let rip with a couple of highly intemperate letters to local gentlemen who had publicly given their support to it. He wrote to the Earl of Strathmore on 10 November 1817, that assigning the invention of the safety lamp to Stephenson was known by "every man of real science in the Kingdom" to be false.[54] The following day, he wrote to the Newcastle barrister James Losh, demanding to know whether Losh, who had been a friend of Davy's friends Coleridge and Southey, was now to be numbered among his enemies. He declared that he had to defend himself against an "attack on my Scientific fame, my honour and varacity [sic]."[55]

Davy was showing himself in a very unfavorable light in these indignant letters. Both Strathmore and Losh responded in a tone of hurt puzzlement, professing their admiration for Davy but insisting that Stephenson also deserved recognition for his accomplishment. Both implied that Davy was behaving arrogantly in belittling an honest man who did not have his social or institutional advantages.[56] Indeed, Davy was mobilizing all the social clout he could command, and he was also threatening to get the Royal Society to take up his cause. As James notes, the dispute pitted the metropolitan knight, savant, and theorist against the humble provincial engineer. An additional resource Davy did not hesitate to deploy was his standing as a philosopher. According himself this title suggested both that his discovery was of a higher intellectual character and that it was motivated by pure benevolence toward humanity at large. Insofar as Davy was recognized as a philosopher, the safety lamp could be represented as a product of his genius and his selfless devotion to human welfare.

Davy's reputation did not emerge unscathed from the safety lamp controversy. His high-handed behavior contributed to the character he was beginning to acquire in some quarters, of an upstart who used his recently acquired wealth and social rank to put down individuals less fortunate than himself. This was to become a repeated theme in the sniping and satire directed against him during the remaining years of his life. On the other hand, the safety lamp became literally the iconic achievement of his efforts to apply science to practical ends. It appears beside him in the portrait commissioned by Lambton from the artist Thomas Phillips and completed in 1821 [Figure 7]; it also features in the painting executed by Sir Thomas Lawrence around the same time [Figure 5, shown in chapter 3]. The presence of the device in these carefully composed portraits shows just how important it was to the public image of Davy's identity. As early as October 1815, Sir Joseph Banks was expressing the hope that the invention would redound to the credit of the Royal Society and characterizing it as an application of "enlightened philosophy."[57] It is plausible to suppose that it helped Davy's campaign to be elected president of that body when Banks died in 1820. The safety lamp seemed to demonstrate that an understanding of fundamental scientific principles could yield discoveries of significant human benefit. It paid testimony to the philosophical benevolence of the man of science. It did not, however, guarantee that further discoveries of equal significance would be made in the future.

Davy's election to the presidency of the Royal Society occurred after the chemist William Hyde Wollaston, who had been serving in an interim capacity since Banks's death, declined to run for the office. Failing to find a viable alternative candidate, Davy's opponents capitulated in the face of his vigorous campaign.[58] Davy himself, writing to Thomas Poole, paid tribute to Wollaston's generosity in stepping aside

Figure 7. Portrait of Humphry Davy by Thomas Phillips, 1821.

This portrait was commissioned by Lord Lambton shortly after he hosted a commemorative banquet in Humphry Davy's honor, in Newcastle in October 1817, to celebrate the invention of the safety lamp. Appropriately, the lamp is shown at Davy's side, as he supposedly composes the paper in which he announced its design. His rather abstracted gaze suggests the contemplative character of a philosopher. The artist, Thomas Phillips, also completed portraits of Michael Faraday and John Dalton. Phillips shared Davy's interests in the techniques used in ancient frescos and in photography. The painting remained in the possession of the Earls of Durham until 1932, when it was acquired by the National Portrait Gallery. It is currently exhibited at Dove Cottage in Grasmere, Cumbria. See Trittel, "Genius on Canvas," 192–95; R. Walker, *Regency Portraits*, 1:148.

"like a true philosopher."[59] The implication was that a philosopher should behave with selfless generosity; the character was incompatible with the open avowal of naked ambition. The question remained, however, whether Davy himself would live up to the ideal. In the course of his presidency, some fellows came to suspect that his own actions showed a degree of ambition and self-interest unfitting for a true philosopher. It was not that the society was hostile to projects that applied scientific knowledge to practical problems. But it was important that those who engaged in them maintain the stance of gentlemanly independence that was central to the society's ethos. Provincial engineers and industrialists who became fellows in this period, including John Smeaton and James Watt, had to present themselves as disinterested philosophers to comply with the institutional norm.[60] Davy was held to the same standard.

A relevant episode was the ill-fated project to protect the hulls of Royal Navy ships from corrosion. Since the 1780s, British naval vessels had been having their hulls sheathed with copper to protect the timbers from the depredations of marine worms. Unfortunately, it was found that the metal quickly became corroded and pitted with holes. Davy was approached for advice on the subject in late 1823 by John Wilson Croker, first secretary of the Admiralty. Davy saw the opportunity to consolidate links between the Royal Society and the government by offering scientific advice on an important issue for overseas commerce and national defense. After a couple of months of characteristically intense experimentation, he proposed a solution in January 1824. Bars of zinc or iron were to be attached to the ships' hulls. According to electrochemical theory, these metallic "protectors" would combine with oxygen in the water, rendering the copper electronegative and preventing its oxidation. While Davy continued a program of experimental trials at the royal dockyards at Chatham and Portsmouth, the Admiralty ordered that the devices should be fitted to seagoing vessels.

As Frank James has shown, the project soon descended into a debacle.[61] Davy joined a voyage of HMS *Comet* to the North Sea and the Baltic in the summer of 1824, to test protectors of various sizes. But already it was beginning to seem that the attachments had the unintended consequence of increasing the infestation of weeds and barnacles on the hulls, the effect of which was to significantly hinder the ships' motion. In January 1825, the *Times* reported that the vessel *Samarang* had returned from North America so infested by weeds as to impede its steerage.[62] Davy, who personally witnessed the ship's return to port, emphatically denied that this was true.[63] But, by the following June, it was clear that several ships had come back from tropical voyages with severely fouled hulls and reduced speed in the water. Croker and other Admiralty officials began to realize the scheme was not going to work. The following month, the navy ordered all protectors removed, and the project was called off. The publicity from the affair was not favorable to Davy's reputation. The *John Bull Magazine* reported that he had kept model ships floating in tanks of seawater outside the Royal Society's premises at Somerset House, where local residents were offended by the smell. They were said to have dubbed the tanks "Numps's ponds," alluding to a disrespectful nickname Davy had acquired.[64] The *Times* judged his experiments a complete failure and declared that the only benefit of the voyage of the *Comet* had been "to himself in procuring a summer excursion, at the public expense."[65] Although Davy had again declined to patent his invention and had announced he was donating it to the nation, he did take advantage of his Admiralty connections to lobby for a naval appointment for one of his brothers-in-law. This appointment led to another fiasco, when the man accidentally killed someone by carelessly discharging a firearm, and his naval career came to an ignominious end.[66]

Facing such antagonism in the press, Davy described his situation as that of a public-spirited philosopher surrounded by petty and vindictive enemies. In October 1824, he wrote to J. G. Children, with whom friendly relations had by this time been restored. He mentioned another "pernicious attack" in the *Morning Chronicle*, behind which, he suspected, stood a group of copper merchants who perceived a threat to their naval supply contracts. "Every body seems to forget," he lamented, "that I have *given* a discovery to the *public*, by which a

great copper manufacturer said He would have made £20,000 a year."
He went on to compare his troubles with those of Galileo, in "the
times when Philosophers & public Benefactors were *burnt* for their
services."[67] Of course, Galileo had not been burned, merely sentenced
to house arrest after his trial by the Inquisition. But the overwrought
comparison reflected Davy's exaggerated sense of his own persecu-
tion. He knew that he ought to be untroubled by the attacks, accepting
them, as it were, philosophically; but he admitted to being more irri-
tated than he should be. Perhaps because he saw his own philosophical
persona as being the target, the matter was not so easily shrugged off.

Indeed, the controversy continued to affect Davy's reputation after
his death. In his chatty and digressive biography, J. A. Paris gave a
lengthy account of the episode of the ships' protectors and concluded
that "the truth of [Davy's] theory was completely established by the
failure of his remedy!"[68] John Davy was obviously dissatisfied with
such a paradoxical assessment and with Paris's apparent receptiveness
to the voices of Davy's critics. His briefer and less nuanced narrative
assigned credit to his brother for discovery of a true scientific principle, while disowning responsibility for the failures to translate it into
practice. "The principle of protection," John Davy claimed, "was perfect."[69] If there were difficulties with the consequent fouling of ships'
hulls, they could easily have been addressed by resourceful naval offi-
cers. Humphry Davy himself had suggested methods for cleaning off
weeds and barnacles, but it was not his duty to carry them out. As
with some of his previous ventures into applied science, the supposedly underlying principles were defined (and distinguished from their
practical applications) in the course of a controversy that followed the
initial innovation. Accounts of how knowledge had been converted
into practice (or not) were set against one another in a lengthy contest
of interpretations by which the meaning of the episode was decided. In
this case, the dispute continued even after Davy's own demise.

Davy's self-image as a philosopher was repeatedly challenged in
the 1820s. At a time of increasing pressure for social and political reform, he was criticized for his loyalty to an aristocratic ideal of gen-

tility and a paternalistic model of practical improvement. His reputation for being in thrall to his social superiors obstructed his attempts to
present himself as a disinterested philosopher. Both inside and outside
the Royal Society, his standing was vulnerable. The short-lived periodical the *Chemist*, published from March 1824 to April 1825, often had
Davy in its sights. The publication was edited by the activist Thomas
Hodgskin and was closely associated with the Mechanics' Institute
movement for working-class education. To these liberal writers, Davy
was representative of an "aristocracy of chemistry," which entrenched
itself in Somerset House, Albemarle Street, or the Athenaeum Club
and excluded all outsiders.[70] Davy was said to have "no appearance
of labouring for the people. He brings not the science which he pur-
sues down to their level; he stands aloof among dignitaries, nobles,
and philosophers."[71] When Davy spoke on the occasion of the unveil-
ing of the monument to James Watt, the authors of the *Chemist* made
a telling comment. They acknowledged Watt's virtues as a man of science who applied his knowledge to practical ends, but they insisted he
had done so for his own material advantage: "The example of Mr. Watt
shows that he is the best citizen who pays the closest attention to his
own interest." Watt's straightforward pursuit of his own ends was contrasted with the stance of the "moral *quacks*" who "pretend to guide
their actions with a view to universal good."[72] Davy's image of himself as a philosopher dedicated to the service of humanity at large had
never been so precisely skewered.

Davy's response to such criticism was to adopt a stance of disdain-
ful superiority and to withdraw increasingly from worldly engage-
ments. He never showed any sympathy for the *Chemist*'s campaign
to educate the working classes. Even when the editors sent him the
first number of the periodical, in which they had been quite complimentary about him, he wrote to Children that he would "never shake
hands with chimney sweepers even when . . . they call me 'your Honour.'"[73] A few years later, he wrote to his wife that he had become increasingly dubious about the project of the Scottish journalist Henry
Brougham to spread scientific knowledge to working people. There

was no use in trying to make ordinary people into philosophers, he declared, since they lacked the Socratic self-knowledge to appreciate their own ignorance. The outcome could only be to encourage religious skepticism and social discontent.[74] Clearly, those who suspected that Davy's notion of himself as a philosopher was tinged with elitism, even a degree of snobbery, were not entirely wrong.

As the 1820s went on, it began to seem that Davy could sustain his self-image as a philosopher only by detaching himself from the affairs of the world. He traveled extensively on the European Continent, especially as his health began to fail. In letters back to his wife, he described himself as a "sick Philosopher" and a "crippled though not subdued philosopher."[75] In his last years, he was composing his *Consolations in Travel*, in which he set out an idealized model of the relations between the "Chemical Philosopher" and the various practical arts. His fantasy self-projection in that text, the character called "the Unknown," represents the way Davy would have liked to be seen: a charismatic individual of unquestioned moral and intellectual authority. The Unknown possesses the profound theoretical knowledge of chemical principles that allows him to grasp their application to all the dependent arts. He is able to present a comprehensive account of all the technical fields to which chemical science has contributed improvements. Intriguingly, despite appearing amid ancient ruins in vaguely classical garb, the Unknown turns out to have some specific attributes that reflected Davy's social situation. The character explains that he "really became a philosopher" only when he attained wealth and freedom to travel.[76] Being a chemical philosopher was evidently not consistent with pursuing science for profit. Only by virtue of financial independence could the Unknown engage in the sublime speculations that would attain the most abstract knowledge while simultaneously satisfying humanity's everyday needs.

As Davy embodied it—and even as he fantasized about it—the figure of the philosopher had an obvious sociological dimension. It encompassed specific features of the gentlemanly identity to which Davy aspired as he climbed the ladder of Regency society. But it would be wrong to reduce the category entirely to these terms. Representing and conceiving himself as a philosopher, Davy was doing more than indulging his social pretensions. He was attaching himself to what he saw as an antique tradition, one he traced in his lectures back to the sages of ancient Greece. He was laying claim to knowledge of a profound intellectual character, with the implication that such knowledge could provide the key to mastery of the material world. And he was laying upon himself a rule of moral conduct, an ethical standard against which he could be judged. The philosopher was expected to place his knowledge at the service of humanity as a whole, not to cultivate personal or sectional ambitions. As his mouthpiece the Unknown explains, "in becoming wiser, he will become better, he will rise at once in the scale of intellectual and moral existence."[77] It is not surprising that such a strenuous ethical ideal should have proven impossible to attain in practice. Nor is it remarkable that Davy's personal moral foibles were on occasion exposed. The ideal was so exacting that it is hard to imagine anyone completely living up to it. And, in the course of Davy's career, the collisions between philosophical ideal and concrete social reality were often bruising indeed.

6. The Traveler

These sublime and magnificent scenes afforded me the greatest
consolation that I was capable of receiving. They elevated me from
all littleness of feeling; and although they did not remove my
grief, they subdued and tranquillized it.

MARY SHELLEY, *Frankenstein*

Humphry Davy was an inveterate traveler. From his boyhood minera-
logical expeditions to the last days of his life, he seemed constantly on
the move. While employed at the Royal Institution, he traveled back to
Cornwall to visit his family, and he also made a geologic survey of his
native county. Other early geologic excursions took him to the Derbyshire Peak District, the Cumbrian Lakes, and North Wales. He first
visited Scotland in 1803, and two years later he ventured to Ireland,
crossing the Irish Sea to observe the basalt formation of the Giant's
Causeway in County Antrim. A more extensive Irish tour followed in
1806, and subsequent visits were built around his lecturing engage-
ments in Dublin.[1] Accepting Napoleon's invitation to visit Paris in the
autumn of 1813 opened up the prospect of Continental destinations.
Accompanied by his wife, Jane, and his younger friend Michael Faraday, he embarked on a lengthy tour to France, Switzerland, and Italy,
eventually coming back to England in May 1815.[2] Three years later,
Humphry and Jane undertook another journey of almost two years, returning to Naples and the buried cities of Pompeii and Herculaneum.[3]

Service as president of the Royal Society [Figure 8] limited his

Figure 8. Portrait of Humphry Davy by James Lonsdale, 1822.

This portrait of Humphry Davy as president of the Royal Society of London is now in the collection of the Royal Society of Edinburgh. It shows Davy apparently restless in the president's chair (previously occupied with great gravity by Sir Joseph Banks), as if eager to resume his travels. It was exhibited at the Royal Academy in 1822, and an engraving was published in 1827. Davy had been elected a fellow of the Royal Society of Edinburgh in 1808, and the painting was acquired by the society in 1849. See R. Walker, *Regency Portraits*, 1:149; http://www.bbc.co.uk/arts/yourpaintings/paintings/sir-humphrey-davy-17781829-bt-frs-frse-186226.

Portrait of Sir Humphry Davy FRSE (1778–1829) by James Lonsdale, reproduced by permission of the Royal Society of Edinburgh.

travels for a while, but when Davy suffered his first stroke at the beginning of 1827, he was prompted again to seek a southern climate that he hoped would restore his health. For several months, he toured the Alps and Italy with his brother, John, who acted as his personal physician.[4] Humphry's final journey was again undertaken with a therapeutic purpose. Accompanied by John James Tobin, the son of an old friend and a student of medicine in Heidelberg, Humphry left England in March 1828 for what turned out to be the last time. He suffered another devastating stroke in Rome in February 1829, and while he clung precariously to life, his wife and brother rushed to join him there. The party began the slow journey home but got only as far as Geneva, where Humphry suffered another—this time fatal—stroke. He was buried in a cemetery just outside the Swiss city, the grave in foreign soil being somehow appropriate to a perennially restless individual.[5]

Humphry Davy's travel was always purposeful; he usually had more than one aim in mind. Writing to the British ambassador to Constantinople during his first Continental journey, to try to secure a permit to visit Greece, Davy declared that he was traveling "principally with a view of applying chemical research to the solution of some important problems in Natural History and to the investigation of some objects connected with the progress of art."[6] Such chemical travels could cover many possible areas of inquiry. During his stay in Paris, he had investigated the newly discovered element iodine, trying to prove to his French colleagues that it spelled the death of Lavoisier's oxygen theory of acidity. In Florence, he burned a diamond to confirm that it was pure carbon in crystalline form. In Rome, he made a chemical analysis of the pigments used in ancient frescoes and dispatched a paper on the subject to the *Philosophical Transactions*.[7] His later attempts to unroll the ancient papyrus scrolls found in the ashes of Herculaneum showed the same willingness to take his chemical expertise on the road and apply it to objects he encountered.[8] While he had access to laboratory facilities in the major cities he visited, most of this work was done with a small collection of instruments housed in two easily transported wooden boxes. The modest apparatus allowed Davy to make his chemical inquiries mobile, moving away from his London base to new sites of experimentation.

Davy's studies of volcanoes combined this sort of laboratory investigation with the methods characteristic of the fieldwork sciences.[9] Volcanoes were themselves chemical laboratories of a kind; Davy referred to Vesuvius as "this grand laboratory of Nature."[10] But they also required study by fieldwork methods of mapping and collecting specimens. Davy tried to prove the igneous origins of basalt, and he believed he had clinched the point when he encountered rocks of a transitional form in the lava fields of the Auvergne.[11] The transition emerged clearly from the specimens he collected as he made his passage across the landscape. Wherever he ventured, he recorded rich descriptions of the space through which he was traveling. He narrated his itineraries in his notebooks, combining comments on the prevailing rock types with remarks on the inhabitants' social customs and economic activities. He also sketched geologic features, using contemporary methods to capture rock formations in diagrammatic form.[12] He often drew landscape views, sometimes labeling the rocks that were exposed. He sometimes sketched the layers of strata that were invisible beneath the surface of the earth, deducing their order from their exposure in a few spots. He also tried to sketch geologic maps, drawing an aerial view of the surface strata, even when they were invisible to view.[13] Davy was not a particularly talented artist, but the standard of execution of his sketches was less important than the purpose they served. They manifested his constant concern to record his movements across a terrain, the spatial loci of the formations he encountered on the route, and what they implied about the structure of the rocks beneath his feet, in the third dimension of the earth's interior.

This kind of travel had an aesthetic aspect as well as a scientific one. Davy frequently recorded his appreciation of picturesque or sublime landscapes. Going beyond emotional responses, he situated himself and his bodily reactions within his accounts of the places he visited.[14] When he started traveling to improve his health, his bodily state began to feature even more prominently. He was convinced that the aesthetic qualities of scenery could have therapeutic benefits, and he was pre-occupied with the relationship between the physical characteristics of climate and his state of health. He painstakingly monitored air temperature and other weather conditions because he was sure they bore upon the heating or cooling of his internal organs, which he understood as the primary cause of his individual pathology.[15] Finding exactly the right climatic conditions was as important to him as the correct diet or the appropriate regime of exercise in the program of therapeutics to which he subjected himself. This was another way in which Davy invoked the classical traditions of care of the self, carefully regulating all the circumstances that impinged on his bodily condition to sustain his physical and mental well-being. He was also aware of the rich history of eighteenth-century deliberations about climate, its bearing on human character and destiny. Davy's therapeutic travel comprised a kind of environmental regimen, a search for the climatic conditions that would be conducive to the recovery of his health.

Many of the experiences of his journeys were reflected in Davy's last work, *Consolations in Travel* (hereafter *Consolations*, 1830), which we have encountered a few times already. Largely dictated during his final tour to Tobin, who served as his amanuensis, the book was published the year after his death. Its contents are remarkably heterogeneous: fragments of autobiography, narratives of dreams, philosophical dialogues concerning religion and immortality, visions of spectral beings and travel to other planets, and disquisitions on chemistry and geology. Read against the background of Davy's lifelong travels and his whole career, it can be recognized as both a sweeping philosophical vision and a deeply personal work.[16] The dialogues involving fictional characters are set in locations that were particularly meaningful to the author: the Coliseum in Rome, the summit of Vesuvius, the temple of Paestum in Campania, the cavern at Adelsberg (now Postojna in Slovenia), and the harbor at Pola (now Pula in Croatia). These were the places that evoked for him the deep questions of the history of life on earth, the future of humanity, and the persistence of the individual spirit after death. They suggested the most profound and meaningful lessons Davy thought could be learned from travel, the consolations it could provide if correctly interpreted.

To understand what these consolations were, we have to examine Davy's self-presentation as a traveler, in this enigmatic final book and in his other writings. We will find that he was recommending and modeling a certain kind of *philosophical travel*. The distinctive features of this mode of travel included a sensibility to particular scenery and an openness to the dimension of time as well as of space. By grasping the workings of time in shaping the landscape, Davy suggested, observers could derive both pleasure and a kind of metaphysical consolation. Uncovering the perspective of the distant past, they would come to appreciate the prospects both for the future of human beings and for personal immortality. There was both an appropriateness and a pathos in Davy advocating this kind of philosophical travel in his final work, composed as it was on the threshold of his own death. Contemplating the final journey of his spirit as it was about to leave his physical body, he seems to have derived comfort from the metaphysical insights travel had brought him.

And yet the character of the traveler does not resolve the conundrum of Davy's multiple identities. *Consolations* remains an ambiguous and puzzling text because it presents several facets of Davy's personality. Even in his last book, he seemed unwilling or unable to settle on just one face to show to posterity. He is present as two or maybe three of the characters, in addition to being identified as the author. The book, the last of a series of experiments in authorship of various textual genres, only obscurely reflects the writer himself. One might say, in fact, that Davy's lifelong trials of identities, his multiple experiments in selfhood, reach their culmination—but not their resolution—in this extraordinary and mysterious work.

On his first day on the European continent, 19 October 1813, in the harbor of Morlaix in Brittany, seasick after a rough Channel crossing, Davy wrote that he was heading south to study "the Chemistry of nature & volcanoes."[17] The phrase captures both the general framework of his travels—an inquiry into nature viewed primarily as the domain of chemical change—and the specific objects that were the main targets of his quest. He was about to fulfill a lifelong ambition to study volcanoes firsthand. He had devoted two lectures to them in his 1805 series on geology, though he had never encountered them in his travels up to that time. Addressing the audience at the Royal Institution, he had been obliged to depend on published descriptions (one of them more than a century old) and on a letter from Etna by his friend G. B. Greenough.[18] Nonetheless, he was clearly captivated by the topic. Greenough's description of the scene of the lava field as "wild, and sublime" was amplified in Davy's own remark that a volcanic eruption must be "beyond all comparison the most awful and the most sublime of the phenomena belonging to our globe."[19] Pondering what the chemical cause of such an eruption could be, he proposed the agency of pyrite (iron sulfide), since this was the "only known substance found in the bosom of our mountains which is capable of spontaneous inflammation."[20] Pyrite, he pointed out, had been observed to ignite spontaneously when it came into contact with water, and so it might be supposed to fuel the intense fires by which volcanoes uplifted and fused rocks. As it happened, the theory survived only until Davy's discovery of the alkali metals sodium and potassium in 1807. That alerted him to the existence of other possible mineral sources of spontaneous ignition, and he soon suggested that the newly discovered metals could be the hidden fuel of volcanic eruptions.

In 1805, Davy showed his audience a model volcano in which pyrite, mixed with coal dust and an oxidizing salt, was ignited by a drop of acid. The demonstration was not his most striking effect—he admitted it was a little "ridiculous"—and it was replaced by a much more dramatic model volcano when the alkali metals became available.[21] In any case, no tabletop demonstration could match the spectacle of the real thing. When he made it to the extinct volcanoes of southern France, and especially when he arrived at the still-active Vesuvius in the spring of 1814, he was transported by the spectacle. Observing the lava fields on the slopes of the mountain and the yellowish fumes emitted from vents in the crater, he recorded that he "never beheld a more magnificent prospect."[22] He had a vivid sense of contact with the most powerful forces of nature, the intense heat that fused and molded minerals into their familiar forms, and the great energy that thrust the mountains up from beneath the surface of the earth. These were forces that deserved to be called sublime. They emerged from the subterranean realm of darkness and disorder, and they conveyed a sense of fearful and intimidating power, before which the human spirit could not help feeling belittled.

There was another dimension to the sublime feelings Davy experienced in his travels. As he made his way, and as he reflected on his journeys in notebooks and letters, he became more acutely aware of the depth of time stretching behind visible phenomena. He declared that volcanic rocks, for example, "have been produced in times of which we have no history or tradition."[23] If one grasped the processes by which minerals and landscape features had been created, it opened up the vista of lengths of time much deeper than human history or even the reach of the imagination. The realization was not unique to Davy at this period. Many other students of the earth's history were coming to grips with an extent of geologic time much greater than their predecessors had admitted, and certainly more than the six thousand years allowed by a literal interpretation of biblical scriptures.[24] To some observers, the sublimity of this expanse of time was associated with the idea that it manifested an extended period of divine creativity. God could be seen to have acted not only at a moment of creation but continuously over the lengthy period during which natural forces played themselves out. To others, such as the followers of the Scottish philosopher James Hutton, the expanded period during which natural forces had acted tended to push God out of the picture. There was no need to invoke divine intervention since there was—in Hutton's notorious phrase— "no vestige of a beginning" in the visible record of the earth's history.[25]

In his 1805 lectures, Davy staked out his position on this contentious topic. He dismissed the kind of geology that remained in thrall to the scriptural timescale. At the same time, he was scornful of the speculative excesses he perceived in the theoretical systems of Hutton and his rivals. Davy invoked the sublime as a kind of veil drawn by the deity across the depths that human knowledge was not permitted to penetrate. He admonished Hutton that "man . . . was not intended to waste his time in guesses concerning what is to take place in infinite duration."[26] He simultaneously castigated Hutton's critic Jean André Deluc for his "vain attempts to penetrate into mysteries that have been wisely concealed from us."[27] The sublimity of the past, Davy suggested, should be respected as an impenetrable mystery. He warned of the dangers of crossing the bounds set for human knowledge and the risks of leading the public astray in pursuit of that goal.

The direct experience of the Alps and the Italian landscape that Davy acquired in his travels made him more inclined to venture across this boundary, though he continued to evoke the mysteries of the deep past that lay beyond. In his private notes, and then in the public prose of *Consolations*, he drew the veil aside, if only part of the way. He voiced his own experience of deep time, which manifested the long-term action of natural laws. There was no suggestion of Hutton's heretical notion of the eternity of the world. But neither did Davy explicitly evoke divine intervention in the course of the earth's long existence. His portrayal of the shaping of landscape by the powerful but slowly acting forces of nature conveyed the sublimity of geologic history, while leaving open the question of God's involvement in the process. Davy was even willing to consider that living creatures might have occurred in different forms over the course of the earth's history, though he was careful to distance these speculations from the discredited materialism of such individuals as Erasmus Darwin.[28] Davy's contemporaries and successors were often inspired by the sublime vistas called up by his consciousness of geologic time, but they were also sometimes frustrated by his ambivalence on the underlying theological issues.

The theme of deep time began to emerge in the course of Davy's first Continental journey. In a series of notes dated May 1814, he recorded his passage from Rome to Naples and the ancient temple at

Paestum.[29] Along with the spectacular sight of Vesuvius, he was struck by the rapid rate of calcareous deposition by the streams descending from the travertine plateau on which the temple sat. Ancient observers had already recorded the petrifying qualities of some of these streams. Geologic processes could be seen unfolding here, overlapping with time's destruction of the remains of human history. Davy noted that ancient Greek artworks at the temple site had survived in better condition than the Roman works, even though the latter were more recent.[30] Historical and geologic changes were also observed at the nearby ruins of Pozzuoli, identified in Davy's time as a temple of Serapis. Like previous visitors, Davy recorded the bands of three or four feet in height around the middle of the temple columns that had been nibbled by marine zoophytes. How could this have happened? Did it mean the temple had been submerged under the sea for a while after its construction and then raised up again onto dry land? What movements of the earth could have caused that? Davy concluded that a more likely scenario was that the columns had been reused from a more ancient temple that had been buried under the sand.[31] As it turned out, he was wrong to dismiss the theory of successive submersion and uplift. Later geologists, including Charles Lyell, who illustrated the columns at Pozzuoli in the frontispiece to the first volume of his *Principles of Geology* (1830), concluded that they did indeed bear witness to substantial changes of elevation and the position of the coastline at this site.[32]

On his next trip to Italy, in 1819, Davy sat down in Bologna to draft an essay on geology which he called, "On the Structure of the Solid Parts of the Globe." Although he had ceased to give public lectures by this date, he adopted the form of a lecture to set out his ideas. He began with the surface of the earth, so wonderfully fitted out as the abode of life in all its variety. He then passed to the interior: obscure and apparently chaotic, the source of the sublime powers by which the surface was molded. He summarized the agency of water and fire in the formation of rocks and the creation of conditions for life. Casting his mind back to the distant past, he concluded that the circumstances in the earliest period of the earth's history were "wholly incompatible with the existence of any organized beings similar to those by which the globe is now peopled."[33] The implication was that entirely different species existed in the conditions prevailing in earlier eras. Davy was reminded of the fact that the remains of marine animals could be found on the tops of mountains, where they obviously could no longer live. It was plausible to conclude that massive deluges had swept over the whole globe in the distant past, just as—on a smaller scale—the English Channel had been inundated more recently. These deluges could be associated with the large-scale changes in conditions that had allowed new forms of life to emerge. Ending with another comparison with ancient (human) history, Davy noted that Plato's story of the submersion of Atlantis might not be the pure myth that was often assumed.

These speculations received confirmation—at least in Davy's view—from investigations in the 1820s by the Oxford professor William Buckland. In 1822, Davy presided over the award to Buckland of the Royal Society's Copley Medal for his inquiries into animal remains found in a cave in Kirkdale, Yorkshire. Davy had visited the site in Buckland's company, and he confirmed the astonishing finding that hyenas, rhinoceroses, hippopotamuses, elephants, and other tropical creatures had once lived in this northern location.[34] In his presidential address, Davy expressed satisfaction that the discovery "establishes, beyond all doubt, the great catastrophe described in the sacred history." Buckland had proved that the history of the earth had to be understood in terms "of a successive creation of living beings, of which man was the last." This was to replace Hutton's idea of the eternity of the world with an episodic and broadly progressivist vision. It was also to assert a theistic position against Enlightenment speculations about the evolution of "organized germs" into fish, quadrupeds, apes, and eventually human beings.[35] Davy later expressed frustration that press reports of his address had made it sound as though he was endorsing biblical literalism as a basis for geology. That was not his point, he explained, though it was often hard to tread a sure path through this highly contested field. The evidence for global inundations did not establish the truth of the biblical story of the Flood, still less its occurrence on a scriptural timescale.[36] The important thing was that Buckland had opened up the vista of a long series of epochs in the history of life on earth, with humans entering at a relatively late stage. Materialist theories of transmutation of species could be dismissed, although the exact cause of the successive creation of different forms of life remained hidden from view.

The sublime vision of deep time Davy had glimpsed in the course of his travels was articulated in *Consolations*. His adoption of the form of a dialogue, with contrasting opinions expressed by the various characters, allowed for the argument to be laid out informally but also left some ambiguity as to the author's own outlook. At the beginning of the book, Davy established the positions of two of his characters in a way that reflected eighteenth-century religious controversies. "Onuphrio" represents the skepticism about traditional doctrine that emerged in the Scottish Enlightenment, while "Ambrosio" speaks from a liberal position within traditional Catholicism.[37] The narrator, "Philalethes," whose views closely match Davy's own, mediates between the two. Thus, in the first scene of the book, these three figures take in the magnificent ruins of the Coliseum and reflect on the legacy of Christianity and classical antiquity. Ambrosio points to the triumph of Christian faith over ancient Roman superstition and its associated barbarities. For him, the most important lesson to be learned from the ruins of Rome concerns the victory of Christianity over its pagan persecutors. Onuphrio, on the other hand, introduces the melancholy reflection that Christian civilization itself is destined to pass away, as that of ancient Rome did, in the long sweep of human history. In the background here were Enlightenment debates on the rise and fall of empires. Davy was invoking such figures as Edward Gibbon, who was inspired to write his account of the decline and fall of the Roman Empire by his own meditations in the Forum in Rome, and the Comte de Volney, whose *Ruins . . . of Empires* achieved great popularity in the 1790s.

The third dialogue brings the conversation around to geology. By this stage, Davy has introduced a fourth character, the mysterious

Unknown, whose authoritative voice contributes a degree of conclusiveness to the debates. He clearly reflects Davy's own views, though the fact that Philalethes is also present—and has already assumed the part of the narrator—created confusion among some readers. The Unknown sets out an account of the history of life on earth. He begins in the earliest era, when fishes, birds, and reptiles were the only living beings, passes to the period when massive now-extinct animals flourished ("the mammoth, megalonix, megatherium, and giant hyena"), and ends at the relatively recent point when "the creation of Man took place."[38] Ambrosio objects that the narrative is not strictly compatible with that presented in the book of Genesis, but he seems satisfied with the reassurance that the Bible was not intended to serve its readers as a scientific textbook. Onuphrio speaks up on behalf of Hutton's system but is gently rebuffed by the Unknown's objections against the idea of an eternal and unchanging order in the past. The Unknown explains that he does not identify the ancient deluges with the biblical Flood, but neither can he accept "that the present order of things is the ancient and constant order of nature, only modified by existing laws."[39] Onuphrio perceives that his own position might be thought to support the materialist writers' claim that new species could evolve by purely natural causation, and he beats a hasty retreat. As he concedes:

> I shall push my arguments no further, for I will not support the sophisms of that school, which supposes that living nature has undergone gradual changes by the effects of its irritabilities and appetencies; that the fish has in millions of generations ripened into the quadruped, and the quadruped into the man; and that the system of life by its own inherent powers has fitted itself to the physical changes in the system of the universe.[40]

On this point, it seems, all the interlocutors agree. What Onuphrio calls "this absurd, vague, atheistical doctrine" is dubbed by the Unknown a "false and feeble philosophy," and on this note of consensus the discussants adjourn for dinner.[41]

Consolations articulated Davy's experiences of the geologic sublime in their most developed form. For him, the deep time of the earth's history evoked emotions similar to those generated by the remains of ancient civilizations. Having witnessed the operation of natural forces, he was deeply impressed with their cumulative power and could imagine their long-term effects in molding the landscape. The reflection provoked melancholy feelings about the passage of time, the shortness of human life, and the inevitable decay of monuments and memorials. It was also clear to him that the processes of time were still somewhat mysterious, that they remained veiled in some degree of obscurity. Davy was resistant to the idea of flattening out the history of the earth by reducing it to uniformity. He rejected Hutton's principle of the expansion of uniform natural laws into the indefinite past. And he spurned the materialist theories of transmutation, associated with Erasmus Darwin and Jean-Baptiste Lamarck, according to which the emergence of new forms of life was explicable by natural and commonplace processes. Deep time was to remain shrouded in obscurity, divided from the present by discontinuous ruptures in the order of things. Its sublimity was associated with its mysteries, with the suggestion—but not the definitive assertion—that God's creative action had occurred outside the normal laws of nature.

In the final dialogue of *Consolations*, "Pola, or Time," Davy reiterated this message. In the harbor at Pola on the Illyrian coast, his characters discuss long-term changes in the material world. They consider the operations of gravity, heat, and chemical forces, the wearing down of mountains by erosion, and the destruction of human monuments by corruption and decay. Faced with the melancholy prospect of the mutability of all things, Philalethes and the Unknown agree that comfort must be sought from faith in God's design. Christianity provides the reassurance that all these transformations are ultimately serving the divine plan. The theme is one that Davy might have taken from his sixth-century precursor Boethius. In Boethius's *The Consolation of Philosophy*, human anxieties and uncertainties are allayed by an awareness of the providential design underpinning the cosmos. Everything must, ultimately, be for the best, since providence rules over the universe as a whole. But the final sentence of Davy's book offers a rather different solace, one that evokes the Roman atheist poet Lucretius more than the Christian Boethius.[42] Speaking at the end, Philalethes suggests that even if human monuments decay to dust, they can sustain the growth of vegetation to nourish new life: "nature asserts her empire over [the ruins] . . . , and the vegetable world rises in constant youth, and in a period of annual successions, by the labours of man, providing food, vitality, and beauty upon the wrecks of monuments which were once raised for purposes of glory, but which are now applied to objects of utility."[43] The vision of the endless recycling of materials through the operations of the laws of nature may offer a kind of consolation, but it also suggests the eternity of matter and the autonomy of its active powers. It is intriguing that, thirty years after Davy penned these lines, Charles Darwin chose a similar image for the conclusion of his book *On the Origin of Species* (1859). Darwin conjured up the same sense of change and eternal renewal with his mention of the "planet . . . cycling on according to the fixed law of gravity" while evolution slowly took its course.[44] It is a testimony to the fundamental ambivalence of Davy's writing—its basic theological convictions obscured by the veil of the sublime—that his image could subsequently be appropriated for a work that explicitly advocated a materialist theory of evolution.

Consolations has always been read as a highly personal book. From the moment of publication, it was immediately and inseparably associated with its recently deceased author. It emerged from his experiences of travel, especially during his last illness. And it comprised his most urgent intellectual legacy, the thoughts of a mind poised on the margin between life and death. And yet, there has always been a degree of uncertainty surrounding Davy's self-representation within the book. His decision to portray himself in two distinct characters, and his apparent intrusions into the text in yet further guises, caused confusion among his readers at the time and ever since. If this is his most personal work, then, it is one that leaves us with a strong sense of the elusiveness of

Davy's personality. Seemingly unable to settle on just one identity, he continued his protean shape-shifting into his final work, and left as his legacy the image of a man who wore many masks and whose real character escaped definition.

Davy began writing the book in June 1828, at Ischl in the Austrian Alps. In his account of the trip, Tobin recorded that Davy was dictating what he called his "Vision" for an hour or two each morning.[45] This dictation turned out to be the episode in the first dialogue, in which Philalethes, after discussing the rise and fall of civilizations with his companions, undergoes an otherworldly experience in the Coliseum. Resting there in the moonlight, he is visited by a mysterious voice, which announces itself as a superior intelligence from another world. The spirit or "Genius" then unveils to the narrator scenes of previous eras in the world's history and life on other planets. In the subsequent chapters, Philalethes and the other characters refer to this vision as they discuss chemistry, geology, immortality, and the history of living things. Davy told Tobin that the episode was based on a dream he had experienced a few years previously, "in which he found himself borne through the firmament from planet to planet." To Tobin, it seemed that Davy's mind was "wandering, as it were, with the associates of his early days," as he conjured up the congenial and brilliant friends who had been the companions of his youth.[46] His mind straying while his body decayed, it seemed that Davy's spirit was already beginning to detach itself from the circumstances of his earthly existence.

While dictating the book to Tobin, Davy was also mentioning it in letters home. In July 1828, he wrote to Jane that he was amusing himself with the composition of what he called a piece of "philosophical poetry, though not in metre."[47] Later that month, he told her that the text would contain "all my philosophical thoughts & poetic feelings."[48] As time went on, he became more convinced of its importance. By December, he was telling Jane, "It may be imagination, but I seem to see the novel scheme of the Universe more clearly than formerly." He attributed the insight to bodily exhaustion, which allowed the mind to become detached from the "passion or enthusiasm" that had previously clouded his judgment.[49] Davy was laying the foundation for the book to be read as the insights of a spirit on the point of freeing itself from the trammels of the physical body and the enthusiastic emotions of youth.

After Davy suffered his stroke the following February, this interpretation was stamped on the work he had just brought to completion. Convinced he had only a few more days to live, he claimed the privilege of a dying prophet, one whose special insight into matters of life and death derived from his position on the boundary between the two states. He decided on the book's subtitle: "the last days of a philosopher."[50] And he charged his brother, John with the solemn duty of ensuring its publication at all costs.[51] In a letter dictated to his wife at the beginning of March 1829, he wrote:

> I should not take so much interest in these works did I not believe that they contained truths which cannot be recovered if they are lost & which I am convinced will be extremely useful both to the moral and intellectual world. I may be mistaken in this point, yet it is the conviction of a man perfectly sane in all the intellectual faculties & looking into futurity with the prophetic aspirations belonging to the last moments of existence.[52]

For anyone who picked the book up after publication, it was impossible to detach it from its author's circumstances at the end of his life. Davy initially planned to be identified on the title-page as "the author of *Salmonia*" (his previous book, on salmon fishing).[53] This was not a serious attempt at anonymity, since it was widely known who that author was, but it would have kept the work at a slight distance from his more mainstream scientific publications. However, when *Consolations* appeared, Davy's name was prominently displayed below the title. Also, the first thing that readers were told, in an introductory note contributed by his brother, was that the book "was concluded at the very moment of the invasion of the Author's last illness."[54] This note was followed by a short Preface, which Humphry Davy himself

had somehow managed to dictate the day after he suffered his debilitating stroke in Rome. In these ways, the book was presented as in-
separable from the individual who had produced it and the circumstances in which he had done so. It owed a large part of its authority to
its association with a philosopher on the threshold of his own death,
or, as a reviewer in the *Dublin Literary Gazette* put it, "when the soul
was quivering on the beam between the two states of existence."[55] The
book was a parting message from a person embarked on his final journey, the one from which no traveler ever returns.

Some of the early reviewers acknowledged that *Consolations* would
not have commanded so much attention or respect had its author lived
to see it published. A writer in the *Athenaeum* concluded that the book
was not without its flaws, but it demanded attention as the final product
of a considerable intellect. The reviewer hailed Davy's "very ingenious,
if somewhat visionary, hypotheses and speculations."[56] The writer in
the *Dublin Literary Gazette* conceded that some of the contents were
"extravagant, and almost bordering on the absurd."[57] A critic in the
Monthly Review was more direct, asserting that "we give to this last
emanation of Sir Humphry Davy's vigorous mind, a degree of attention, which its own abstract claims may not be sufficient to justify."[58]
The work was vulnerable to criticism, this reviewer said, on grounds
of metaphysical abstraction, ponderous prose, and a chaotic structure.
Some of the obituary notices for Davy ignored *Consolations* altogether,
reflecting the sense that it was an anomaly among his publications and
a bit of an embarrassment.[59] But, whatever its peculiarities, the book
proved popular, and that popularity was sustained for the remainder
of the nineteenth century. There were at least nine individual editions
published in London before 1900 and two more in which *Consolations*
was combined with other works by Davy. An American edition, published in Philadelphia, followed hard on the heels of the first London
one, and another appeared in Boston forty years later. Translations in
German, Swedish, and Dutch were published within a few years of
the book's first appearance, and other languages followed.[60] A literary
monster it might have been, but there is no doubt that it drew signifi-

cant numbers of readers, and continued to draw them in subsequent decades.

One of the difficulties the text posed for readers was the question of where exactly Davy stood in the dialogues. While the author's name was clearly printed on the title page, and the book was universally known to have emerged in the last days of his life, it was not straightforward to locate him and his views in the text itself. *Consolations* flouted the normal conventions of written dialogues or conversations, in which the author was identified with one particular position. Davy didn't appear in his own person, and he seemed to have more than one representative among the fictional participants. Philalethes is the narrator for parts of the text, and the vision in the Coliseum is assigned to him, though a few pages later he admits it was a composite of dreams he had had on various occasions.[61] The effect here is the rather dizzying one of the author dropping his character's mask or jumping the frame of the narrative. Although still speaking in the voice of Philalethes, Davy seems to be peeping out from behind him, admitting that his character's vision was fictional, while insisting it was based on a dream *he* had really had.

The conversation then moves on to another dream of special significance for Davy. Again, it is Philalethes who tells the story of how, in the delirium caused by a serious illness, he saw a vision of a beautiful woman then unknown to him. Ten years later, he was reminded of the vision when he encountered the attractive young daughter of an innkeeper in Illyria. "Now," says Philalethes, "comes the extraordinary part of the narrative: ten years after, twenty years after my first illness, at a time when I was exceedingly weak from a severe and dangerous malady, . . . I again met the person who was the representative of my visionary female; and to her kindness and care I believe I owe what remains to me of existence."[62] It may be Philalethes who is speaking in the text, but the dream was Davy's own, as was the insistence that it had truly been prophetic. The young woman who was his muse and later his nurse was Josephine Dettela, from Laybach (now Ljubljana in Slovenia). Davy wrote a poem about her in August 1827, using the pet name "Pepina." He celebrated her "virgin purity" while calling on her to kiss him.[63] The relationship had a flirtatious air, though Davy idealized it as a platonic alliance of two souls and held fast to the notion that it had been foretold in his dream. When the connection is alluded to in *Consolations*, there is a strange oscillation between author and character, a motion of masking and unmasking on Davy's behalf. He assigned the significant dream to his character Philalethes. And yet, to guarantee its validity as prophecy, he had to drop the mask and repossess it for himself, insisting it was not fictional but a real event.

John Davy addressed the puzzlement all this caused for readers, when he wrote his brother's biography in the late 1830s. He claimed that the Coliseum vision was based on a dream Humphry had had in Rome in 1819.[64] The other incident was dated to earlier in Humphry's life, to 1807, when he had experienced a vivid and apparently prophetic dream during his bout of typhus.[65] On the other hand, John explained that his brother had never fallen down a waterfall or traveled to Palestine, although these things happened to characters in the book.[66] The confusion was not surprising, and it was compounded by the introduction of the character of the Unknown in the third dialogue. As the Unknown begins to give his opinions—about the possible virtues of chlorine as a preventive against malaria, about the geology of the region, and eventually about the history of the planet and living things— readers seem again to hear Humphry Davy's own voice. In the fourth dialogue, the Unknown is encountered once more in the Austrian Alps, where he rescues Philalethes from nearly drowning in a boating accident. The near-fatal experience is followed by a discussion on personal immortality and whether life continues after death. In the next dialogue, the Unknown assumes the role of the chemical philosopher, voicing what is clearly Davy's own claim to that title. The Unknown personifies this unworldly figure, who by virtue of his abstraction from mundane desires could attain true wisdom and realize the true utility of science.

Readers were nonetheless puzzled by Davy's having introduced another character to represent himself when Philalethes already seemed to be playing that role. Accustomed to the conventions whereby an author spoke unequivocally through one character, they were confused by Davy's adoption of more than one persona. The *Monthly Review* thought that the Unknown was a more authentic mask for Davy to assume, "notwithstanding that the author appears to have disposed of his own identity already in the character of Philalethes."[67] John Davy suggested that his brother took several roles in the book because he was himself a man of many parts: poet, metaphysician, geologist, chemist, and Christian. Underlying all of them was the figure of the philosopher, "in the original, modest, and humble meaning of the word."[68] The Unknown was therefore the most fundamental of Humphry's identities. John recorded that he had discussed this matter with his brother and had evidently encountered some resistance to the suggestion that the Unknown was another self-portrait. John ventured, nonetheless, to contradict his brother and insisted that the identification was correct. John wrote: "independent of his dress and some of the incidents of his life, he *was* essentially the prototype, in sentiments, feelings, opinions, doctrines, — in brief, in mind; . . . The religious sentiments *The Unknown* expresses, and his metaphysical doctrines, were, I believe, entirely my brother's own."[69]

The matter of religious sentiments posed a further conundrum for Humphry Davy's readers. He had always been publicly identified as a member of the Church of England, and yet *Consolations* seemed unexpectedly pro-Catholic in its sympathies. Ambrosio, who upheld the side of religious orthodoxy in the jousts with Onuphrio, was explicitly identified as a Catholic and was soon recognized as being based on a distinguished ecclesiastic who had been Davy's host in Italy.[70] When the Unknown comes on the scene, there is a curious incident in which Ambrosio identifies him also as a Catholic because he is wearing a rosary. The Unknown explains that he is actually an Anglican but that he wears the rosary in tribute to Pope Pius VII, whom he met in exile at Fontainebleau.[71] Again, readers assumed that the episode had happened to Davy himself. The reviewer in the *Athenaeum* wrote that the anecdote was "so pleasing, that we would be sorry to think it is not a literal account of a real occurrence; and it has more value also in our eyes when we persuade ourselves that we may consider the narrator of the story as Sir Humphry Davy himself."[72] It was again left to John Davy to try to sort the matter out in his biography. He noted that the story has the Unknown receiving the rosary at the Church of the Holy Sepulcher in Jerusalem, but Humphry Davy never actually visited the Holy Land. He *was* in Rome in 1814, when Pius VII returned to the city from captivity under Napoleon, but there is no evidence that he actually met the pontiff. It seems that Davy did have a degree of respect for Catholicism and for at least some of its devotees, but readers who thought he was endorsing Catholic doctrine were misled by the slippage between the author and his characters. With the constant masking and unmasking in which Davy indulged in *Consolations*, the confusion is understandable.

Given what we know about Davy's lifelong propensity for experimenting with his own identity, his introduction of more than one figure to represent himself in this book is not so surprising. As we have seen, he was always adopting masks of one kind or another, in his public performances as a scientific lecturer, in making his rapid rise up the social ladder, and in his writings. *Consolations* could be seen as his last virtuoso performance, the last of his experiments in selfhood. He also seems to have believed that a fictional character, not too closely identified with himself, possessed the kind of charisma needed to convey his message. In one of his notebooks in 1827, he had made some jottings intended to introduce some sort of autobiography. It would, he remarked, focus not on himself but on "a very extraordinary Person . . . a mind far superior to my own." Of this individual, he wrote: "His opinions are wont to be so singular[,] his lessons so instructive & his history so mysterious that they are worthy of being recorded."[73] By the end of these notes, Davy had dubbed this individual "the Unknown"; he had discovered the character who was later to assume such a prominent role in *Consolations*. Davy noted that the character would seem forever youthful, like one who had discovered the elixir of the alchemists or the secret of eternal life.[74] With these attributes, he would compel readers to attend to the message he had to impart.

So the Unknown was conceived as an unworldly—even other-worldly—being, not so much a flesh-and-blood person as a spirit or angel temporarily assuming human form. This allowed the character to speak with authority on a pivotal theme of the book: the prospect of intellectual or spiritual life continuing beyond the death of the material body. This was obviously of urgent concern for Davy at the time. Afflicted by bodily ailments, he sought comfort in the idea that mind and body were not inseparably connected, that the former could survive independently of the latter. He repeatedly reassured others (and himself) that his sickness was affecting only his body, that his mental capacities remained unimpaired. The play with multiple masks and roles was a way to reinforce the point that the intellect or spirit was not to be identified with an individual's physical manifestation. Bodily features change in the course of someone's life, but intellectual identity remains. This was what gave grounds for the hope that it might continue in some form after death.

The point is made in a rather unexpected way in the fourth dialogue, titled "The Proteus, or Immortality." After Philalethes is rescued from nearly drowning, he and a companion called "Eubathes" go with the Unknown into the cavern at Adelsberg. Philalethes is musing on his near encounter with death, but the immediate object of the journey is to study the amphibian, *Proteus anguinus*, which dwells in the cave. These organisms had drawn Davy's attention in the last months of his life, and the conversation touches on the peculiar features revealed by his inquiry. The creatures are able to live below or above water, on the surface of the rock or deep in the mud, because they possess both gills and lungs. Their origins are obscure, as are their feeding habits and mode of reproduction. The discussion then passes to the chemistry of respiration, and whether the air could be the source of a material principle that constitutes vitality. The Unknown is adamant that the mystery of life cannot be explained in this materialistic manner. "I can never believe . . . that *intelligence* can result from combinations of insensate or brute atoms," he insists.[75] Mental processes cannot be identified with the material organs of the senses. On the contrary, the mind is an immaterial entity that exists continuously throughout the changes in a person's body during his or her lifetime. And at the end, as the Unknown puts it, "the mind, as it were, falls asleep, to awake to a new existence."[76]

Although the rationalist Eubathes resists this conclusion, Philalethes professes himself satisfied with the Unknown's deduction. Speaking again for Davy himself, Philalethes denounces materialism, as "a cold, heavy, dull and insupportable doctrine . . . necessarily tending to atheism."[77] He neglects to say that he had in fact flirted with it in his youth, during his association with Thomas Beddoes and other radical thinkers. He soon changed his mind, and by the 1810s he had openly allied himself with the antimaterialist side of a bitterly fought controversy that was dividing the scientific—and especially the medical—community.[78] The reviewers had no difficulty discerning which side he was taking in this dispute. The *Monthly Review* welcomed his attack on the "sophisms" of materialist physiologists, the *British Magazine* expressed its appreciation for his support of the orthodox Christian doctrine, and the *Anti-Infidel* hailed his demolition of a theory "which some half-informed venders of blasphemy are so anxious to diffuse."[79] On this issue, it seems, Davy had managed to speak without equivocation, having aligned the positions of both of his spokesmen on the antimaterialist side of the debate.

The proteus was something of a test case for materialist theories about the nature of life. In the dialogue, the Unknown discusses but rejects the suggestion that it is an immature form, like a caterpillar or a tadpole, of some creature that looks quite different as an adult. In his private notes, Davy had raised a more radical possibility: that the proteus might be transitioning into another species altogether. He wrote, "It gives the idea of a sort of preformation for a more perfect animal as if a germ of creation & it is impossible to contemplate it without singular speculations concerning the return of the mysterious process by which every part of the globe has been peopled with beings fitted to exist in them."[80] In the published text of *Consolations*, the suggestion is advanced tentatively by Philalethes, who claims to have been "carried in imagination back to the primitive state of the globe."[81] He compares the anatomy of the proteus with that of the "sauri," the giant lizards that were beginning to be recognized as inhabitants of an earlier phase of the earth's history. The implication is that the little amphibian might be a newly created form, destined to give rise to other creatures by some process of transmutation. Philalethes apologizes for the speculation, "which I suppose you will condemn as wholly visionary and unphilosophical," and is gently reproved by Eubathes for having made remarks "unworthy of discussion."[82] But the idea had been given an airing. In a rather similar manner, Davy had speculatively raised the possibility in *Salmonia* that one kind of fish might be capable of changing into another by the inheritance of acquired characteristics. Facing criticism from reviewers, he was obliged to insert clarifying comments into the second edition in which he emphatically distanced himself from the "unsound" theories of Erasmus Darwin concerning the transmutation of species.[83]

In *Consolations*, the mention of the possible transmutation of the proteus is followed by a discussion of the survival of the human soul after death. The characters ponder the persistence of personal selfhood throughout life and after the demise of the physical body. These had long been among Davy's preoccupations, and they were given additional urgency by his precarious health at the time of writing. On the last pages of the book, he invoked the theory of the astronomer William Herschel, who claimed that stars and planets could be formed from the nebulae glimpsed through telescopes. Davy even suggested that such planetary systems could be inhabited by "genii or seraphic intelligences."[84] Presumably the "Genius" who had appeared to Philalethes in the Coliseum was of this species; and perhaps the souls of the departed would be destined to join them. Davy himself may have hoped to escape from his mundane life into what Tobin called his "oft self-imagined planetary world."[85] Whatever the exact form of the transition he was hoping for, Davy was obviously anticipating some kind of continuity of mental existence after death. The humble—but appropriately named—proteus was an apt symbol of this expected transformation.

One might say that the proteus was the last of Davy's self-projections in *Consolations*. Alongside Philalethes and the Unknown, the cave-dwelling creature may well have reflected something of himself. Its amphibiousness mirrored his own mobility, his extraordinary adaptability to different institutions and social circles. Its pervasiveness and the mysteries of its origin echoed his own circumstances. Even the enigma of its reproduction had parallels in his life experience. In a letter to his wife, Davy compared the writing of *Consolations* to bringing a child into the world; he may well have been musing about his own failure to father real children as his life neared its close.[86] As the shape-shifting god, Proteus had featured prominently in William Wordsworth's sonnet "The World Is Too Much with Us," composed around 1802. Davy had also reflected in one of his notebooks that the mythical being served as a symbol for the transmutations of matter studied in the science of chemistry.[87] Now, at the end of his life, he was looking for signs of a transformation that would go beyond the conversion of one kind of matter into another. What he was seeking, poised on the brink of the mysterious gulf between mortal and spiritual lives, was the separation of the immaterial soul from its material substrate. For this transition, a simple amphibian offered itself as a fitting emblem.

Epilogue
A Fragmented Legacy

John James Tobin was right to suggest that Humphry Davy's last days saw him revert to some of the preoccupations of his youth. There was more than a hint of self-experimentation in the way he carefully monitored and adjusted his therapeutic regime during his travels. His brother, John, and other physicians had told him that his temperament was a sanguine one, "in which there is a tendency to excess of sensibility and irritability, and of vital action, combined with corresponding activity of mind, and a certain warmth and impetuosity of temper."[1] Aware that his problems were caused by a warm constitution, he adjusted his diet to avoid rich food and drink, and he took care not to expose himself to hot climates. He called this the "antiphlogestic" [*sic*] regimen, an ironic label in view of his criticisms of French antiphlogistic chemistry.[2] Believing that his strokes had been brought on by rapid flows of blood to the head, he applied leeches to his neck and temples and had himself bled frequently, especially when he felt a headache coming on.[3] As his brother acknowledged, people who did not understand the rationale for these measures might think the frantic changes of location and treatment bespoke "derangement of intellect"; but they were in fact reflections of Davy's "activity of mind and unyielding disposition."[4] They were, in other words, not pathological symptoms but the signs of an active genius who was continuing to take himself as the subject of experimental inquiry.

If this recalled his early trials of gaseous therapies, Davy was also reviving his old interest in living things. He studied his leeches as well as using them to draw blood, examining the lives of the little creatures on which his own life depended. Writing to Davies Giddy from Ravenna in March 1827, he reported that the leeches he had brought from London had frozen in the course of the passage through the Alps, but once thawed out, they had bitten immediately.[5] At the same time, he extended his physiological inquiries into other living organisms. He began the study of *Proteus anguinus* in the cavern of Adelsberg and brought to conclusion a lengthy series of experiments on the torpedo, or electric fish. These investigations, pursued in various Mediterranean locations over more than a decade, formed the subject of his last paper in the *Philosophical Transactions* in October 1828.[6] While noting that his poor health would prevent him from pursuing the inquiry any further, he typically did not hesitate to use his body as an instrument to measure the intensity of the shocks the fish could produce.[7] It seems the torpedo was an appropriate way to close the circle of Davy's lifelong study of the powers underlying biological processes, in which he had often resorted to taxing experiments on his own body.

His work on the proteus provided an outlet for these interests. As I have suggested, Davy could see aspects of himself reflected in the puzzling little amphibian. The possibility that it was an immature form of an animal that would look quite different as an adult spoke to Davy's anxieties over his impending death and the somber transition it promised. The enigma of the organism's mode of reproduction would also have captivated a man without biological children and worried about his legacy. And the question of the proteus's place in an evolutionary process can also be asked of Davy himself. If the animal was a transitional form, undergoing transmutation into some other creature entirely, then one might wonder what the goal of the alteration was. Evolution, as Davy understood it, was a directed process, oriented toward some designed end, not the outcome of blind material forces. So, if the proteus stood for Davy's own place in a sequence of historical changes, where was the transition heading? What exactly was the longer-term trend in which he featured?

The question is easy to ask but not at all easy to answer. And perhaps the temptation to ask it should be resisted. I have been arguing that Davy's identity was in many respects a fragmented one. He deployed multiple personae, depending on the circumstances in which he found himself, while making his selfhood a topic of serious experimental inquiry. It should not be surprising, then, that his successors tended to perceive him in a fragmentary manner. As early as 1821, the literary essayist William Hazlitt mentioned the difficulty of assessing his character as a whole. Davy was, Hazlitt acknowledged, "a great chemist, but I am not sure that he is a great man."[8] Overall judgment of his reputation was further complicated by changes in the scientific community toward the end of his lifetime. Later men of science appropriated aspects of his work and modeled themselves on features of his life that appealed to them, while ignoring or downplaying the rest. For this reason, it has never been easy to situate Davy in a narrative of the emergence of the specialist man of science or the professional scientist.

As an author, for one thing, he bequeathed to his successors a heterogeneous body of work. The eight-volume *Collected Works* (1839–40), which his brother compiled as an act of fraternal devotion, gave a misleading impression of uniformity. Therein, Davy's immature essay on light and life nestles alongside the researches on pneumatic medicine that made his name. His landmark Bakerian lectures on the alkali metals are accompanied by more ephemeral speeches, fragments, and extracts from his journals. The first part of the *Elements of Chemical Philosophy* is reprinted, although the project was aborted after this initial phase. The widely read *Agricultural Chemistry* is spread across two volumes, so as not to damage sales of the edition already on the market. And the durable *Consolations in Travel* shares a volume with the much less popular *Salmonia*. Publication of such a collection was a rare tribute for a British man of science in the nineteenth century, though it created only a superficial kind of unity among Davy's varied publications. The sense persisted of a man who was exploring many authorial personae in the different textual genres that flowed from his pen.

Much of this virtuosity—the simultaneous appeal to multiple audiences in many authorial guises—came to seem amateurish and suspect to Davy's successors. Leading figures among his immediate posterity were hostile to many aspects of his scientific character. In the decade after his death, a coalition appeared that redirected British science toward completely new institutional forms. Charles Babbage, the Cambridge mathematician who had spearheaded opposition to Davy's rule in the Royal Society, emerged as the prophet of a new movement to reverse what he claimed was the "decline" of science in England. John Herschel, astronomer and meteorologist, was another Cambridge graduate whose zeal for reform had been frustrated under Davy's presidency. The "Gentlemen of Science," comprising Oxbridge specialists, Edinburgh savants, and well-placed provincial intellectuals, coalesced around formation of the British Association for the Advancement of Science in 1831.[9]

In the same period, a trend toward the foundation of specialized scientific societies was gathering steam. Chemists, geologists, astronomers, and others began to establish their own institutions—a movement toward independence from the Royal Society that Sir Joseph Banks had squashed in his day. Institutionalization was accompanied by the development of new outlets for scientific publication, especially cheaper books and more frequently published periodicals. William Thomas Brande, Davy's successor as professor of chemistry at the Royal Institution, rode the wave of some of these innovations. He became a leading member of the Chemical Society of London when it was founded in the 1840s, and he served as the long-standing editor of the *Quarterly Journal of Science*. He was looked down on by Davy, who in private notes referred to him as "a very inferior person" who "followed chemistry always (as a German apothecary might be expected to do) . . . [for] as much profit as He could obtain."[10] Brande, who knew that Davy himself had been apprenticed to an apothecary as a boy, resented his air of superiority. In 1826, he wrote to John Murray, publisher of the *Quarterly Journal*, noting Davy's complaint about a paper it had published and advising Murray to "pay no kind of attention to this exceedingly impertinent interference of that self-constituted autocrat of science."[11] Obviously, relations between the two men were strained, but the spat also reflected the changing circumstances Davy was encountering in his last decade. The journal the *Chemist* was also critical of his "undue pretentions" and "jealousy of others" in his role as president of the Royal Society.[12] In an age of mounting pressure for political and institutional reform, Davy's attempt to continue Banks's learned empire was interpreted as a vain and arbitrary autocracy.

In these circumstances, many of Davy's traits of character came to be viewed unfavorably. The artful rhetoric and deliberate displays of passion by which his reputation for genius had been created began to seem insincere, even a little histrionic. Subsequent writers on scientific method emphasized steady application and the accumulation of facts, rather than leaps of intuition. Davy's reliance on the patronage of Whig magnates and the landowning aristocracy also looked in retrospect rather demeaning. His cultivation of an audience of elite admirers, including many women, seemed indiscriminate. As the gentlemen of science staked out their careers in later decades, they drew firmer lines between the domain of specialist communication and that of popularization. Women were invited to attend open sessions of such groups as the British Association for the Advancement of Science, when discoveries were presented in attractive and accessible language; but they were not allowed to trespass on discussions among specialists at the frontiers of research, which were confined to men only.[13] Davy's personae as discoverer and philosopher were also soon eclipsed as models for scientific behavior. By authoring a textbook in the exclusive character of a discoverer, he was held to have given far too much prominence to his own findings and opinions. Davy might have despised Brande as a man without significant discoveries to his name, but the latter wrote a more successful textbook precisely because he was not so biased toward his own laboratory results.[14] Nor did Brande or most of his contemporaries mimic Davy in clinging to the character of a philosopher. As disciplinary specialization advanced, "man of science" and eventually "scientist" became the preferred monikers. The rhetorical maneuvers by which Davy had tried to secure credit for practical innovations, while disclaiming—as a philosopher—any desire to profit from them, looked in retrospect like hypocrisy.

This is not to say that Davy lacked influence on his successors. They might have defined features of their scientific personae by differentiating themselves from him, but they could not have achieved what they did without him. Subsequent men of science distanced themselves from what they saw as the vain pursuit of fashion or female adulation, but they continued to court public audiences for their work. The market for popular science continued to flourish, and scientific specialists continued to operate in it. In fact, some of Davy's fiercest critics at the end of his life, including Babbage and Herschel, were also writing books aimed directly at general readers.[15] Davy's identification with the electrical battery—his recruitment of it as a prosthetic extension of his own bodily powers—might have seemed self-indulgent or self-promoting, but its legacy can be discerned in later developments. Users of subsequent contraptions, including electrical, magnetic, and optical devices, mobilized the same natural powers to serve a similar aesthetics of visual display.[16] Davy's particular kind of "romantic science" emerged in the specific settings of Bristol and London in the early 1800s, but its echo reverberated through the remainder of the nineteenth century.[17] Throughout Europe, Davy was recognized as a worthy peer of the like-minded savants with whom he corresponded, including A. M. Ampère, H. C. Ørsted, and J. J. Berzelius. His status as a leading contributor to the science of his era cannot be disputed, any more than the lasting legacy of his discoveries can be denied.

Davy's character as a traveler was also reinterpreted, as *Consolations in Travel* continued to be widely read. Later scientific travelers perused this work alongside those of his German contemporary Alexander von Humboldt. During his extensive journeys in South America during the period 1799–1804, Humboldt had assembled verbal descriptions and a huge corpus of quantitative data from astronomical, meteorologic, magnetic, and geologic instruments. His massive publications, appearing gradually over the following three decades, included stunning maps and innovative diagrammatic representations. In this respect, he went much further than Davy, whose book was de-

void of visual images. But the two writers shared an interest in recording their bodily and emotional reactions to landscape as they moved
through it. Humboldt consulted instruments and feelings in tandem,
as Davy had, using one to calibrate the other. Davy would have agreed
with him that "descriptions of nature affect us more or less power-
fully, in proportion as they harmonize with the condition of our own
feelings."[18]

Unsurprisingly, then, the geologist Charles Lyell, who was a keen
reader of Humboldt, was also among the first to seize upon Davy's *Consolations*. He quoted a lengthy passage from Davy in the first volume
of his *Principles of Geology*, published the same year (1830). He was
particularly impressed by Davy's ability to discern the effects of long-
acting natural forces in shaping the landscape, approving his description of the sedimentation of travertine marble in springs and lakes in
Campania. On the other hand, Lyell resisted Davy's supposition that
evolution manifested a progression of forms of life. At this point in his
career, Lyell regarded such theories as unproven and speculative. He
rather condescendingly referred to Davy as "a philosopher . . . pleased
to indulge in conjectures on this subject."[19] When it suited him, Lyell
invoked the negative connotations of Davy's self-description as a
philosopher—the sense of a loosely speculative, possibly unreliable
thinker.

Within ten years, Charles Darwin had also become a reader of *Consolations*. Darwin packed Humboldt in his kit during his own travels
in South America and got down to Davy's book shortly after his return to England. He read it just as he was beginning to formulate his
own ideas about evolution. Davy had compared philosophers' ignorance of the causes of life to that of a "savage" contemplating a steam
engine. In a telling remark, Darwin recorded that such a person might
be more impressed by a piece of colored glass than by a steam engine.
The comment seems to reflect not only Darwin's actual experience of
the people he called "savages" during the voyage of the *Beagle* but also
his increasing sympathy for materialism.[20] His point was that imputa-
tions of design are relative to human capacities: if we don't have the
relevant experience, we cannot even recognize the steam engine as

something that was designed. The implication is that one should hesitate before identifying divine design in the universe at large or in individual creatures.

Both of these thinkers read Davy against the direction of his argument, but both were captivated by his prose and by the image of the author it conveyed. Lyell disputed the notion of the progressive evolution of living things, but he valued Davy's insight into the long-term effects of continuous forces in geology. Darwin tested his own emerging materialism against Davy's antimaterialist perspective, but he did so by seizing on a simile Davy had proposed. *Consolations* showed both men—and many others—how one could draw from an experience of landscape a profound sense of the history of the earth. In this way, the book displayed the character of a philosophical traveler, someone who thought deeply about the meaning of the world through which he journeyed and who felt its beauty and sublimity in his inmost being. Lyell and Darwin were both looking for ways to model their own selves as scientific travelers. They found inspiration in Davy, notwithstanding the important differences in intellectual outlook.

In other respects, also, aspects of Davy's character were reinterpreted to make them more acceptable in changing times. In his classic work of 1859, Samuel Smiles cited Davy as an example of *Self-Help*. Smiles celebrated his rise from the Penzance apothecary's shop, which Davy himself had rather preferred to forget. Ignoring the chemist's dependence on patronage, he memorialized him as an entirely self-made man. He portrayed Davy as humble, meticulous, and industrious—a laborious and highly disciplined thinker without any distinctive native talents, who was contrasted with the chaotic genius of his good friend Coleridge.[21] This was a travesty. Davy had prided himself on his imaginative flights of genius and his ability to tap into his emotions. He could be methodical in pursuit of a line of inquiry, but nobody who had looked into his notebooks could think of him as rigidly disciplined in his thinking. Smiles made up the Davy he wanted, for his own purposes, and in doing so created a model of Victorian self-improvement. In a similar way in the following century, J. G. Crowther hailed Davy as a hero of the rising working classes. The Marxist journalist, who was fascinated by Davy's physical attributes, wrote that he was "more important sociologically than any of his contemporaries" because his social ascent made him "the chief prophet of the new class of applied scientists."[22] Davy would have been appalled by such a verdict.

There were many reasons, then, why Davy's legacy to his scientific successors was problematic. His final years as an institutional administrator left unpleasant memories among those who soon seized control of British science. Many of his fellow chemists resented what they took to be his social snobbery. Even those who claimed to admire him often seemed to misunderstand him, including some of the most sympathetic readers of his last book. With all of these indications of a fragmented intellectual legacy, we can see why Davy's successors found it difficult to see him whole. Because Davy was always something of an enigma to his contemporaries, his reputation was further obscured by the passage of time. Was he a brilliant discoverer or vain dandy? Technical genius or lofty and disinterested philosopher? Enthusiastic self-experimenter or meditative traveler? The contradictions between the diverse aspects of his character were accentuated as the years passed. In particular, it became harder to locate him within the historical narratives of specialization and professionalization that soon came to be imposed on his period. Davy never seems to have wanted to become a specialist or a professional, in line with what later became the norm. His attachment to institutions and the organs of state was always mediated by relations of patronage rather than by impersonal regulations. And his take on the scientific method emphasized intuition and passionate enthusiasm rather than rationality or dogged industry. In all these respects, he just didn't fit into what became the accepted story of the development of scientific institutions and identities.

Perhaps the only appropriate verdict is the one decided on by Davy's characters in *Consolations*, as they contemplate the mysterious proteus. Debating whether the creature is an immature or transitional form, on the way to becoming something else but maybe halted on its path, they decide it is none of these but in fact "a perfect animal of a peculiar species."[23] That seems like the right way to describe Humphry Davy too.

Acknowledgments

Anyone who writes a biographical study will inevitably be prompted to reflect on his or her own life and the debts accumulated along the way. It is a pleasure to acknowledge my own, amassed over the years this project has been in gestation. Davy featured in my first book, *Science as Public Culture* (1992) and in an article published a few years later ("Humphry Davy's Sexual Chemistry," *Configurations* 7 [1999]: 15–41). The primary stimulus to return to him and write the present book came from my participation in the Davy Letters Project, which is working to publish all his surviving correspondence. I must therefore first pay tribute to Tim Fulford and Sharon Ruston, the leaders of the project, and to Frank James and David Knight, my colleagues as advisory editors. Sharon first extended the invitation to join the team and also arranged for me to spend a month at the University of Salford in the summer of 2009 to work on Davy's letters. Tim and David have given generously of their time in reading my manuscript and commenting on it. Frank has done the same and also pulled me away from the Royal Institution Archives for several stimulating and congenial lunches. Through Sharon and Frank, I also had the pleasure of meeting three doctoral students—Iain Watts, Wahida Amin, and Hattie Lloyd—who are carrying Davy studies forward into the next generation. Each of these people has his or her own view of Davy, and none of them should be held responsible for the shortcomings of mine, but

I have been greatly encouraged by their willingness to consider my interpretation and the example of their own work on the individual we all find so fascinating.

I was first able to conceive and outline this book during the 2008–9 academic year, when I was privileged to hold the Dibner Distinguished Fellowship at the Huntington Library in San Marino, California. I owe a great debt to Roy Ritchie, then director of research at the Huntington, for the offer of the fellowship and his generous hospitality in the course of the year. The friendship extended by members of the Huntington staff and my cofellows made that time a particularly productive and enjoyable one for me. I was subsequently honored to receive a Gordon Cain Distinguished Fellowship at the Chemical Heritage Foundation in Philadelphia in the fall of 2012. I particularly appreciate the friendship and support of Carin Berkowitz, director of the Beckman Center, and Ron Brashear, director of the Othmer Library. Along with all of their colleagues at CHF, they made my time there very rewarding. With the aid of a sabbatical leave from my home institution, the University of New Hampshire, I was able to make substantial progress on the book during the following spring semester. I thank Eliga Gould, chair of the Department of History at UNH, for his support throughout this project, and Kenneth Fuld, dean of the College of Liberal Arts, for providing a research grant that covered some of the associated travel expenses.

I am grateful to Jane Harrison and her colleagues for making it possible for me to spend many hours reading Davy's manuscripts in the archives of the Royal Institution. I also appreciate the assistance of librarians and archivists at the following institutions for facilitating access to manuscripts, books, and images: Huntington Library, Othmer Library of the Chemical Heritage Foundation, British Library, American Philosophical Society, Royal Society of London, Royal Society of Edinburgh, National Portrait Gallery, Academy of Natural Sciences of Philadelphia, and Dimond Library of the University of New Hampshire.

I was delighted to be invited to participate in the conference on

"Romantic Disorder: Predisciplinarity and the Divisions of Knowledge 1750-1850," at Birkbeck College, University of London, in June 2009. Conversations on that occasion with James Chandler, Jon Klancher, Adriana Craciun, Luisa Calè, and others encouraged me to develop the project and to publish an outline of the overall argument ("Humphry Davy: The Experimental Self," *Eighteenth-Century Studies* 45 [2011]: 15-28). Subsequent presentations were given at the University of Salford, Chemical Heritage Foundation, Columbia University, Harvard University, University of Toronto, University of Barcelona, and the Three Societies Joint Meeting in Philadelphia. The honor of receiving the Lindberg Award for Excellence in Teaching and Research of the College of Liberal Arts allowed me to try out an early version of chapter 6 on colleagues and students at the University of New Hampshire. I am grateful to the audiences at all of these presentations for their reception and the stimulus of their questions.

Karen Darling at the University of Chicago Press has been a model editor, encouraging and gently nudging this project along from its earliest days. Evan White and the editors of the Synthesis series have also been very supportive, and the three anonymous referees commissioned by the press supplied conscientious and helpful reports. Audra Wolfe went way beyond what could have been expected from a member of the Synthesis editorial board; she read the whole manuscript thoroughly and offered hundreds of suggestions for stylistic and substantive improvements. The book has benefited greatly from her scrupulous editorial eye.

Invariably left to the end on these occasions, but always most heartfelt, are the thanks due to family and friends. To my family—to Tricia and Simon, Bob and Jane, Eleanor and Frances—I offer deep gratitude for your love and support, especially during a time of family bereavement. To the following friends, I tender my thanks for encouragement, conversation, companionship, and good humor: Carin Berkowitz, Nadine Berrini, Paola Bertucci, Jeff Bolster, James Delbourgo, Nick Dew, Patricia Fara, Michael Ferber, Charlie Forcey, Bob Frank, David Frankfurter, Lige Gould, Jo Guldi, Nicky Gullace, Rebecca Herzig,

Adrian Johns, Sarah Lowengard, Lisa MacFarlane, Anna Maerker, Janet Polasky, Lissa Roberts, Julia Rodriguez, Neil Safier, Lucy Salyer, Simon Schaffer, Anne Secord, Jim Secord, Dipankar Sen, Steven Shapin, Mary Terrall, Jake Viebrock, Heidi Voskuhl, Simon Werrett, Bob Westman, Alison Winter, Audra Wolfe, and Sara Wolper.

Notes

INTRODUCTION

1. Charles Bell to John Richardson, 18 August 1839, in Bell, *Letters of Sir Charles Bell*, 369. My thanks to Carin Berkowitz for bringing this passage to my attention.

2. Shelley, *Frankenstein*, 180. The epigraphs of the following chapters are taken from these pages of the 1818 edition: 33 (Chapter 1), 22 (Chapter 2), 37 (Chapter 3), 34 (Chapter 4), 181 (Chapter 5), and 74 (Chapter 6).

3. Treneer, *Mercurial Chemist*.

4. Lord Glenbervie, quoted in Miller, "Between Hostile Camps," 29.

5. Ross, "Scientist."

6. For professionalization, see Morrell, "Professionalisation." For reservations about the applicability of the concept in this period, see R. Porter, "Gentlemen and Geology"; Lucier, "Professional and Scientist in Nineteenth-Century America"; Desmond, "Redefining the X Axis." On the history of disciplines, see Kuhn, "Mathematical versus Experimental Traditions"; Chandler, "Introduction: Doctrines, Disciplines, Discourses, Departments."

7. W. Walker, *Memoirs of Distinguished Men of Science of Great Britain*.

8. [Brewster,] "Memoirs of the Life of Sir Humphry Davy," 101.

9. See MacLeod, "Distinguished Men of Science."

10. [Brewster,] "Memoirs of the Life of Sir Humphry Davy," 102.

11. Brush, *History of Modern Science*; Cunningham and Williams, "De-Centring the 'Big Picture.'"

12. For that period, see especially Morrell and Thackray, *Gentlemen of Science*; Yeo, *Defining Science*; Snyder, *Philosophical Breakfast Club*. Note that although the word "scientist" was coined in the 1830s, it did not come into common use until much later; its absence in Walker's 1862 print is telling.

13. Miller, "'Into the Valley of Darkness'"; Gascoigne, *Joseph Banks and English Enlightenment*.

14. For self-fashioning, see Greenblatt, *Renaissance Self-Fashioning*. The theme has been used by several historians of science, including Biagioli, *Galileo Courtier*; Shapin, *Social History of Truth*; Shapin, "The Image of the Man of Science." For the self-fashioning of authors in print, see Biagioli and Galison, *Scientific Authorship*; Johns, *Nature of the Book*; Sher, *Enlightenment and the Book*.

15. Shapin, *Scientific Life*.

16. The supposition that there is such an ethos was central to the sociology of science pioneered by Robert Merton. See Merton, *Sociology of Science*; Ben-David, *Scientist's Role in Society*.

17. [Brewster,] "Memoirs of the Life of Sir Humphry Davy," 103.

18. Examples include Sennett, *Fall of Public Man*; Taylor, *Sources of the Self*. On this topic, see also Smith, "Self-Reflection and the Self"; Reiss, *Mirages of the Selfe*; R. Martin and Baressi, *Naturalization of the Soul*; Metzner, *Crescendo of the Virtuoso*.

19. The argument is made particularly forcefully in Wahrman, *Making of the Modern Self*.

20. See, in particular, Jordanova, "Sex and Gender"; Jordanova, *Sexual Visions*; Schiebinger, *The Mind Has No Sex?*; Laqueur, *Making Sex*. Also relevant are Cohen, "'Manners' Make the Man; Keller, *Reflections on Gender and Science*.

21. Wheeler, *Complexion of Race*.

22. A pioneering work was Foucault, *History of Sexuality*, esp. vols. 2 and 3. See also Golinski, "Care of Self and Masculine Birth of Science"; Corneanu, *Regimens of the Mind*; M. Jones, *Good Life in Scientific Revolution*.

23. Humphry Davy to John Davy, 23 February 1829, Davy Letters Project.

24. Biographical works include Paris, *Life of Sir Humphry Davy*; J. Davy, *Memoirs of the Life of Sir Humphry Davy*; Treneer, *Mercurial Chemist*; Thorpe,

Humphry Davy; Hartley, *Humphry Davy*; Knight, *Humphry Davy*; Fullmer, *Young Humphry Davy*; Lamont-Brown, *Humphry Davy*. For comments on the biographies, see Fullmer, "Davy's Biographers." For biblio-
graphical information, see Fullmer, *Sir Humphry Davy's Published Works*.

25. For Watt, see MacLeod, "James Watt, Heroic Invention"; Miller, *James
Watt, Chemist*. For Faraday, see Cantor, *Michael Faraday, Sandemanian
and Scientist*; James, *Michael Faraday*; Cantor, "Scientist as Hero." On
biography in the period, see also Yeo, "Genius, Method, and Morality";
Higgitt, "Discriminating Days?"

26. Browne, *Charles Darwin*; Desmond and Moore, *Darwin*; Hodges, *Alan
Turing*; Mialet, *Hawking Incorporated*. Other biographical studies that
have influenced me include White, *Thomas Huxley*; T. Porter, *Karl Pearson*; Pancaldi, *Volta*.

27. For reflections on the genre of scientific biography, see Shortland and
Yeo, *Telling Lives*; T. Porter, "Is the Life of a Scientist a Scientific Unit?";
Terrall, "Biography as Cultural History of Science"; Daston and Sibum,
"Scientific Personae"; Söderqvist, "'No Genre of History Fell under More
Odium than that of Biography.'"

28. Fullmer, *Young Humphry Davy*, 172–79.

29. Royal Institution MSS HD/20/a, 123.

CHAPTER 1

1. H. Davy, *Collected Works*, 1:53.

2. Previous treatments of the topic include: Fullmer, *Young Humphry Davy*,
211–24; R. Holmes, *Age of Wonder*, 235–304; Jay, *Atmosphere of Heaven*,
169–99; Golinski, *Science as Public Culture*, 153–87. See also Jay, "Atmosphere of Heaven"; Cartwright, *English Pioneers of Anaesthesia*; Colás,
"Aesthetics vs Anaesthetic"; Lefebure, "Humphry Davy"; Hoover, "Coleridge, Humphry Davy, and Early Experiments"; Jacob and Sauter, "Why
Not Pain-Alleviating Effects?"

3. Golinski, *Science as Public Culture*, 105–17; Miller and Levere, "'Inhale It
and See?'"; Dolan, "Conservative Politicians, Radical Philosophers."

4. H. Davy, *Collected Works*, 1:53.

5. J. Davy, *Memoirs of the Life of Sir Humphry Davy*, 1:101.

6. On the connection, see Levere, "Dr. Thomas Beddoes and Lunar Society"; Miller and Levere, "Inhale It and See?'" For background on the Lunar Society, see Uglow, *Lunar Men.*

7. H. Davy, *Collected Works*, 2:74.

8. Humphry Davy to Mrs. Grace Davy, 11 October 1798, Davy Letters Project.

9. Jay, "Atmosphere of Heaven," 304.

10. H. Davy, *Researches Chemical and Philosophical*, 343.

11. Ibid., 518, 531.

12. Ibid., 494.

13. Ibid., 495.

14. Ibid., 496.

15. Ibid., 515.

16. Ibid., 457–58.

17. Cartwright, *English Pioneers of Anaesthesia*, 108, 112–14; Stansfield and Stansfield, *Thomas Beddoes, M.D.*, 280, 295; Fullmer, *Young Humphry Davy*, 76–78.

18. H. Davy, *Researches Chemical and Philosophical*, 542–43n.

19. Humphry Davy to Henry Penneck, 26 January 1799, Davy Letters Project; Fullmer, *Young Humphry Davy*, 273–83.

20. H. Davy, *Researches Chemical and Philosophical*, 462, 491.

21. Ibid., 552.

22. Ruston, *Creating Romanticism*, 132–74.

23. H. Davy, *Researches Chemical and Philosophical*, 553n.

24. Sarafianos, "Contractility of Burke's Sublime"; Vermeir and Deckard, *Science of Sensibility.*

25. H. Davy, *Researches Chemical and Philosophical*, 501.

26. See especially Jay, "Atmosphere of Heaven"; Lefebure, "Humphry Davy."

27. Jacob and Sauter, "Why Not Pain-Alleviating Effects?"; Cartwright, *English Pioneers of Anaesthesia.*

28. Colás, "Aesthetics vs. Anaesthetic."

29. For these protocols, see Shapin, "Pump and Circumstances"; Shapin and Schaffer, *Leviathan and the Air-Pump*, 22–79; Bazerman, *Shaping Written Knowledge*; Haraway, "Modest Witness."

30. Golinski, "Care of Self and Masculine Birth of Science."

31. Schaffer, "Self Evidence."

32. Strickland, "Ideology of Self-Knowledge"; Stewart, "His Majesty's Subjects"; Steigerwald, "Subject as Instrument."

33. For Beddoes's politics, see Porter, *Doctor of Society*. The Edinburgh phrenologist George Combe drew a materialist conclusion from the effects of nitrous oxide in his *Letter to Francis Jeffrey*, 29.

34. Royal Institution MSS: HD/13/f, 47.

35. Royal Institution MSS: HD/13/f, 33, 41.

36. The suggestion is made in Jay, "Atmosphere of Heaven." See also Ruston, *Creating Romanticism*; Richardson, *British Romanticism*, 51–53.

37. H. Davy, *Researches Chemical and Philosophical*, 488–89 (emphasis in the original).

38. Humphry Davy to [James Webbe Tobin], 21 March 1800, Davy Letters Project (emphasis in the original). The letter, held in the Beinecke Library at Yale University, can be dated from the postmark. The addressee can be identified from the destination: Tobin's London lodgings at Barnard's Inn, Holborn.

39. On this issue as it arose in Locke's philosophy, see Taylor, *Sources of the Self*, 159–76.

40. Royal Institution MSS: HD/20/a, 109–12, 121–31.

41. Royal Institution MSS: HD/20/a, 130. The thought was one that gave Davy comfort later in his life, when contemplating his own demise, as we shall see in chapter 6.

42. Royal Institution MSS: HD/21/b, 4. The theme of the fetal environment and life before birth was often treated by Davy's literary friends. Wordsworth, for example, touched upon it in his 1802 ode "Intimations of Immortality from Recollections of Early Childhood,", as did Coleridge in his sonnets on the birth of his son Hartley in 1796. (I thank Tim Fulford for these references.) On the general theme of the milieu in which life develops, see also Mitchell, *Experimental Life*, 144–89.

43. H. Davy, *Researches Chemical and Philosophical*, 528.

44. Ibid., xii–xiii.

45. On mesmerism, see Darnton, *Mesmerism and End of Enlightenment*; Winter, *Mesmerized*; Schaffer, "Astrological Roots of Mesmerism"; Fulford, "Conducting the Vital Fluid."

46. On electrical therapies, see Bertucci, "Revealing Sparks"; Bertucci, "Elec-

trical Body of Knowledge"; Elliott, "'More Subtle than the Electric Aura'"; Barry, "Piety and the Patient."

47. R. Porter, "Sexual Politics of James Graham"; Otto, "Performing the Resurrection."

48. Paton-Williams, *Katterfelto*.

49. Belcher, *Intellectual Electricity*.

50. Delbourgo, "Common Sense, Useful Knowledge."

51. Haygarth, *Of the Imagination*.

52. Humphry Davy to Davies Giddy, 3 July 1800, Davy Letters Project.

53. Royal Institution MSS: HD/20/b, 163.

54. Quoted in Hoover, "Coleridge, Humphry Davy, and Early Experiments," 15.

55. Davy to [Tobin], 21 March 1800.

56. H. Davy, *Researches Chemical and Philosophical*, 459.

57. Royal Institution MSS: HD/20/b, 115, 107, 157.

58. On this topic, see Heyd, "The Reaction to Enthusiasm"; Klein, "Sociability, Solitude, and Enthusiasm"; Mee, "Anxieties of Enthusiasm"; Mee, *Romanticism, Enthusiasm, and Regulation*; and Pocock, "Enthusiasm."

59. Royal Institution MSS: HD/13/e, 3–14, at 11.

60. Beddoes, *Notice of Some Observations*, 4–5.

61. Ferriar, review of *Notice of Some Observations*.

62. *Anti-Jacobin Review and Magazine* 6, no. 23 (May 1800): 109–18, at 109.

63. *Anti-Jacobin Review and Magazine* 6, no. 26 (August 1800): 424–28, at 425, 426.

64. Humphry Davy to John Tonkin, 12 January 1801, Davy Letters Project.

65. *Chemist* 1, no. 9 (1824): 133–35, at 135.

66. Paris, *Life of Sir Humphry Davy*, 65.

67. James Watt to Joseph Black, 9 November 1799, in Robinson and McKie, *Partners in Science*, 309–11, at 310.

68. Wright, "Davy Comes to America."

69. Quoted in Jacob and Sauter, "Why Not Pain-Alleviating Effects?" 171–72.

70. Macnish, *Anatomy of Drunkenness*, 88.

71. Paris, *Life of Sir Humphry Davy*, 65.

72. [Anon.,] "Sir Humphry Davy, Bart.," 45.

73. Holland, *Journal of Elizabeth Lady Holland*, 61.

74. Humphry Davy to John King, 22 June 1801, Davy Letters Project.

75. *Philosophical Magazine* 10 (1801): 86–87.

76. Dalhousie University Archives, James Dinwiddie MSS: B.76, 13 December 1808.

77. Dalhousie University Archives, James Dinwiddie MSS: E.6, 28 January 1809; E.7, 28 March 1810.

78. Humphry Davy to John King, 14 November 1801, Davy Letters Project.

79. Royal Institution MSS: HD/20/b, 153, 182.

80. Cottle, *Reminiscences*, 199.

81. Ibid.

82. H. Davy, *Researches Chemical and Philosophical*, 468–71. In 1811, the young chemist Amédée Berthollet, son of Claude Louis Berthollet, committed suicide by breathing carbon monoxide, writing down his reactions and feelings until the very end. For a description of the episode, see Thomson, *History of Chemistry*, 2:151.

83. Royal Institution MSS: HD/20/a, 133.

84. Royal Institution MSS: HD/13/j, 150.

85. Herzig, *Suffering for Science*.

86. J. Davy, Memoirs of the Life of Sir Humphry Davy, 1:253–56.

87. Royal Institution MSS: HD/26/D/36 (comments extracted from the eleventh edition of Henry, *Elements of Experimental Chemistry* [1829]).

CHAPTER 2

1. Humphry Davy to Davies Giddy, 8 March 1801, Davy Letters Project.

2. Ibid.

3. On the construction of the figure of the genius in this period, see Yeo, "Genius, Method, and Morality"; Schaffer, "Genius in Romantic Natural Philosophy"; Fara, *Newton*. For the application to Davy, see Forgan, *Science and Sons of Genius*; Kenyon, "Science and Celebrity"; and, for general background, see McMahon, *Divine Fury*.

4. Humphry Davy quoted in J. Davy, *Memoirs of the Life of Sir Humphry Davy*, 1:35.

5. Martineau, *History of England during Thirty Years' Peace, 1849–50*, 1:594.

6. Simond, *Journal of a Tour*, 2:151.

7. Ibid., 1:42–43.

8. [Anon.], "Life of Sir Humphry Davy," 371.

9. Ibid.

10. [Anon.], "Sir Humphry Davy, Bart.," 62.

11. "Discourse Introductory to a Course of Lectures on Chemistry," 2:311–26, at 318.

12. [Anon.], "Life of Sir Humphry Davy," 371.

13. Simond, *Journal of a Tour*, 1:43.

14. Ibid., 2:150–51.

15. Studies of the early history of the Royal Institution include H. Jones, *Royal Institution*; Berman, *Social Change and Scientific Organization*; James and Peers, "Constructing Space for Science"; Knight, "Establishing the Royal Institution"; T. Martin, "Origins of the Royal Institution"; T. Martin, "Presidential Address"; and Unwin and Unwin, "Humphry Davy and Royal Institution." On the general institutional context, see Klancher, *Transfiguring Arts and Sciences*, 44–48, 51–84; Lightman, "Refashioning Spaces of London Science."

16. Quoted in James and Peers, "Constructing Space for Science," 141.

17. Humphry Davy to Grace Davy, 31 January 1801, Davy Letters Project.

18. Davy was expected to engage in analyses of soils, manures, minerals, and ores at the Royal Institution, for which he was paid separately from his salary. See Berman, *Social Change and Scientific Organization*, 56–66. The issue of how he reconciled this kind of work with his self-conception as a philosopher will be taken up in chapter 5.

19. Russell, *Science and Social Change*, 151–52.

20. Chilton and Coley, "The Laboratories of the Royal Institution in the Nineteenth Century"; James and Peers, "Constructing Space for Science."

21. Dibdin, *Reminiscences of a Literary Life*, 226.

22. H. Davy, *Collected Works* 1:88; report in *Philosophical Magazine* 10 (1801): 86–87.

23. Dalhousie University Archives, James Dinwiddie MSS: B.76, 17 December 1808.

24. Unwin and Unwin, "Humphry Davy and Royal Institution," 18.

25. The comment is recorded by Simond, *Journal of a Tour*, 1:44.

26. Silliman, *Journal of Travels*, 2:211.

27. Dibdin, *Reminiscences of a Literary Life*, 226.

28. On the London lecturing scene, see Hays, "London Lecturing Empire"; Morus, Schaffer, and Secord, "Scientific London"; Inkster, "Science and Society in the Metropolis"; Inkster, "Public Lecture as Instrument of Science Education." For the wider scene of scientific popularization, see Cooter and Pumfrey, "Separate Spheres and Public Places"; Fyfe and Lightman, "Science in the Marketplace."

29. On the shows and spectacles of the era, see Baer, *Theatre and Disorder*; Bermingham, "Urbanity and Spectacle of Art"; Morus, *Frankenstein's Children*; Altick, *Shows of London*; Brewer, "Sensibility and the Urban Panorama."

30. On these two institutions, see Inkster, "Science and Society in the Metropolis"; Stewart and Weindling, "Philosophical Threads."

31. Dalhousie University Archives, James Dinwiddie MSS: B.75, B.76, B.77, B.79, B.81, C.53.

32. On metropolitan establishments that competed with the Royal Institution, see Klancher, *Transfiguring Arts and Sciences*, 66–72.

33. [Anon.,] "Life of Sir Humphry Davy," 371.

34. Ibid., 372.

35. Babbage, *Reflections on Decline of Science*, 188.

36. H. Davy, *Collected Works*, 2:315, 318.

37. Unwin and Unwin, "'Devotion to Experimental Sciences.'"

38. Dibdin, *Reminiscences of a Literary Life*, 226.

39. Dalhousie University Archives, James Dinwiddie MSS: E.9, 23 February 1811.

40. [Brewster,] "Memoirs of the Life of Sir Humphry Davy," 103.

41. For Brewster's interest in scientific genius, see *Oxford Dictionary of National Biography*, s.v. "Brewster, David," accessed 2 April 2015, http://www.oxforddnb.com; Yeo, "Genius, Method, and Morality."

42. [Brewster,] "Memoirs of the Life of Sir Humphry Davy," 103.

43. Davy and Banks had envisioned the new society as an informal dining club, and Davy resigned when it became clear that most of the members

wanted to present papers on their research. See Humphry Davy to George Greenough, [November 1807?], Davy Letters Project; Davies, *Whatever Is Under the Earth*, 10–13; Knight, "Chemists Get Down to Earth."

44. Unwin and Unwin, "Humphry Davy and Royal Institution."

45. J. Davy, Memoirs of the Life of Sir Humphry Davy, 1:255.

46. Ibid., 1:255.

47. Cottle, *Reminiscences*, 198.

48. Humphry Davy to Grace Davy, 31 January 1801, Davy Letters Project (emphasis in the original).

49. Levere, "Dr. Thomas Beddoes at Oxford."

50. Thomas Poole was reported to have shared this concern with Humphry Davy. See J. Davy, *Fragmentary Remains*, 62.

51. Samuel Taylor Coleridge to Robert Southey, 21 October 1801, in Griggs, *Collected Letters of Samuel Taylor Coleridge*, 2:767.

52. Samuel Taylor Coleridge to Samuel Purkis, 17 February 1803, in Griggs, ed., *Collected Letters of Coleridge*, 2:927.

53. [Anon.,] "Life of Sir Humphry Davy," 371. A similar comment occurs in [Anon.,] "Sir Humphry Davy, Bart.," 47–48.

54. Humphry Davy to John King, 22 June 1801, Davy Letters Project.

55. Humphry Davy to John King, 14 November 1801, Davy Letters Project. The letter has been dated by the postmark on the original, held in the Bristol Record Office. The passage in which Davy admits to having succumbed to vice was censored in the transcript published in J. Davy, *Fragmentary Remains*, 65–66, and preserved in Royal Institution MSS HD/26/D/45.

56. [Brougham,] review of Davy's First Bakerian Lecture, *Edinburgh Review* (January 1808), 390.

57. [Brougham,] review of Davy's Second Bakerian Lecture, *Edinburgh Review* (July 1808), 399.

58. Brougham, "Davy," 110.

59. [Anon.,] "Life of Sir Humphry Davy," 371–72.

60. Humphry Davy quoted in J. Davy, *Memoirs of the Life of Sir Humphry Davy*, 1:34.

61. Knight, *Humphry Davy*, 122.

62. Faraday, *Curiosity Perfectly Satisfyed*, esp. 144–52.

63. Humphry Davy to John Davy, 15 October 1811, Davy Letters Project.

64. Humphry Davy to John Davy, 22 April 1812, Davy Letters Project.

65. Humphry Davy to Jane Davy, 30 June 1827, Davy Letters Project.

66. Humphry Davy to Jane Davy, 1 September 1828, Davy Letters Project.

67. Leonard Horner to Alexander Marcet, 10 April 1821, National Library of Scotland MS 9818, f.88–89. I thank Frank James for supplying a transcript of this letter, which is also cited in Fullmer, "Humphry Davy's Adversaries," 154.

68. Miller, "Between Hostile Camps," 36. For an alternative account of Davy's presidency, see Fullmer, "Humphry Davy, Reformer"; and for a different view of the society in the period, see Hall, *All Scientists Now*.

69. Because the other secretaryship was held by John Herschel, appointment of Babbage would have conferred both offices on Cambridge mathematicians, a situation Davy clearly thought would tilt too far toward the reform group.

70. Edward Ryan to Charles Babbage, 24 November 1826, quoted in Miller, "Between Hostile Camps," 39.

71. Miller, "Between Hostile Camps," 38; Babbage, *Reflections on Decline of Science*. On Babbage and his tract, see also Secord, *Visions of Science*, 52–79.

72. Miller, "Between Hostile Camps," 36.

73. Lord Glenbervie's diary, 20 January 1817, quoted in Miller, "Between Hostile Camps," 29. On Glenbervie's interest in his own genealogy, see *Oxford Dictionary of National Biography*, s.v. "Douglas, Sylvester, Baron Glenbervie," accessed 2 April 2015, http://www.oxforddnb.com.

74. [Anon.,] "Humbugs of the Age," *John Bull Magazine* (1824), 89. Ollapod was a character in George Colman's 1801 play *The Poor Gentleman*, a village apothecary and would-be gentleman, a sportsman and a flirtatious scoundrel.

75. Miller, "Between Hostile Camps," 23.

CHAPTER 3

1. Simond, *Journal of a Tour*, 1:43.

2. See, for example, Dalhousie University Archives, James Dinwiddie MSS: E.7, 19 February 1810, and E.11, 9 February 1811.

3. [Southey,] *Letters from England*, 167.

4. William Buckland to Roderick Murchison, 27 March 1832, quoted in Morrell and Thackray, *Gentlemen of Science*, 150.

5. On the consolidation of gender identities and roles in this period, see M. Cohen, *Fashioning Masculinity*; Daston, "Naturalized Female Intellect"; Jordanova, "Sex and Gender"; Knott and Taylor, *Women, Gender, and Enlightenment*; Schiebinger, *The Mind Has No Sex?*

6. Soper, "Feminism and Enlightenment Legacies," 710.

7. On the climate of homophobia in the period, see Crompton, *Byron and Greek Love*, esp. chap. 1; Gilbert, "Sexual Deviance and Disaster." On the scandals surrounding alleged homosexual behavior by leading politicians, see McCalman, *Radical Underworld*, chap. 10.

8. H. Davy, *Collected Works*, 8:354.

9. Maria Edgeworth to Humphry Davy, 1 December 1806 and 21 January 1810, Royal Institution MSS: HD/26/D/30 and HD/26/D/31. On the Edgeworths' interest in education, see Uglow, *Lunar Men*, 315-16.

10. Studies of this literature include Douglas, "Popular Science and Representation of Women"; Sutton, *Science for a Polite Society*; Mazzotti, "Newton for Ladies."

11. Macaulay, Letters on Education; Wollstonecraft, *Vindication of the Rights of Woman*; Polwhele, Unsex'd Females.

12. *Times*, 19 October 1813, 3.

13. H. Davy, *Collected Works*, 2:325, 326.

14. Leonard Horner to J. A. Murray, 15 November 1804, quoted in H. Jones, *Royal Institution*, 264.

15. [Marcet,] *Conversations on Chemistry*. On Marcet, see also Myers, "Jane Marcet's *Conversations on Chemistry*"; Bahar, "Jane Marcet and the Limits to Public Science"; Knight, "Accomplishment or Dogma"; Lindee, "American Career of Jane Marcet's *Conversations on Chemistry*"; Polkinghorn, *Jane Marcet*; Rossotti, introduction to Chemistry in the Schoolroom.

16. [Marcet,] *Conversations on Chemistry*, 2-3.

17. Ibid., 2.

18. Humphry Davy to Alexander Marcet, 29 May 1806, Davy Letters Project.

19. Humphry Davy to Alexander Marcet, 28 October 1810, Davy Letters Project.

20. Simond, *Journal of a Tour*, 2:151.

21. Eleanor Anne Porden to Mary Ann Flaxman, 11 April 1812, Derbyshire Record Office, Matlock, DD3311/25/1/17. I thank Adeline Johns-Putra for supplying this reference. See also Johns-Putra, "'Blending Science with Literature.'"

22. Royal Institution MSS: HD/15/f, 3.

23. For background on changing models of masculinity in this culture, see Carter, "Tears and the Man"; M. Cohen, *Fashioning Masculinity*; M. Cohen, "'Manners' Make the Man"; Fulford, *Romanticism and Masculinity*; Hitchcock and M. Cohen, *English Masculinities*; Hunt and Jacob, "Affective Revolution."

24. Paine, *Rights of Man*, 80.

25. *Times*, 19 October 1813, 3.

26. J. Davy, *Memoirs of the Life of Sir Humphry Davy*, 1:136.

27. Ibid., 2:387.

28. [Anon.,] "Humbugs of the Age," 90.

29. Ibid., 89, 91.

30. Scott, *Journal of Sir Walter Scott*, 107, 109.

31. Carlyle, *Sartor Resartus*, 204.

32. Moers, *Dandy*, esp. chaps. 1, 2.

33. [Anon.,] "Sir Humphry Davy," *Mirror of Literature*, 63.

34. Carlyle, *Sartor Resartus*, 205; Adams, "Hero as Spectacle."

35. [Anon.,] "Sir Humphrey Davy and His Visit to Paris," 675. See also *Times*, 19 October 1813, 3; *Times*, 5 April 1814, 3.

36. "I. S.," letter to the editor, *Times*, 21 October 1813, 2.

37. Paris, *Life of Sir Humphry Davy*, 268.

38. John Davy to Thomas Poole, 28 May 1831, Royal Institution MSS: HD/26/D/54. Michael Faraday, who was accompanying Davy on this journey, recorded his own displeasure at seeing the looted works in the Louvre but did not comment on Davy's attitude. See Faraday, *Curiosity Perfectly Satisfyed*, 15.

39. Moers, *Dandy*, 36–37.

40. This is discussed in Hunt and Jacob, "Affective Revolution."

41. Crompton, *Byron and Greek Love*, 12–62.

42. Samuel Taylor Coleridge to Humphry Davy, 3 February 1801, in Coleridge, *Collected Letters*, 2:671.

43. Levere, *Poetry Realized in Nature*, 20–35.

44. Samuel Taylor Coleridge to Robert Southey, 13 January 1804, in Coleridge, *Collected Letters*, 2:1028. In letters from Bristol in 1799, Southey had described Davy as "a miraculous young man," whose discovery of nitrous oxide had invented a new pleasure and who was destined to achieve the first rank among chemists. See Robert Southey to William Taylor, 21 February 1799, 12 March 1799, and 5 September 1799, in Huntington Library, San Marino, California, MSS HM 4819, 4820, and 4823.

45. Samuel Taylor Coleridge to Robert Southey, 14 December 1807, in Coleridge, *Collected Letters*, 3:41–43.

46. Samuel Taylor Coleridge to Robert Southey, 21 October 1801, in Coleridge, *Collected Letters*, 2:768. As several critics have pointed out, the eponymous hero of Mary Shelley's *Frankenstein* (1818) shares the fate that Coleridge worried Davy would succumb to: cutting himself off from the love of a woman by directing his energies toward the service of humanity in general. See, for example, Mellor, *Mary Shelley*, esp. chap. 6.

47. Samuel Taylor Coleridge to Robert Southey, 21 October 1801, in Coleridge, *Collected Letters*, 2:768.

48. Humphry Davy to John King, 14 November 1801, Davy Letters Project.

49. Royal Institution MSS: HD/14/i, 70.

50. Anna Beddoes to Humphry Davy, 26 December 1804, in Royal Institution MSS: HD/16/H/9.

51. See *Oxford Dictionary of National Biography.*, s.v. "Giddy, Davies," accessed 2 April 2015, http://www.oxforddnb.com.

52. See, for example, Royal Institution MSS: HD/20/a; and HD/15/j.

53. [Anon.,] "Humbugs of the Age," 91. On the reputation of the Edinburgh Bluestockings, see Rendall, "'Women That Would Plague Me.'"

54. Jane Apreece [Davy] to Walter Scott, 4 March 1811, Davy Letters Project.

55. Bury, *Diary of Lady-in-Waiting*, 1:54.

56. Scott, *Journal of Sir Walter Scott*, 108–9.

57. Fullmer, "Humphry Davy and Gunpowder Manufactory," 178.

58. Faraday to Benjamin Abbott, 25 January 1815, in Faraday, *Curiosity Perfectly Satisfyed*, 146.

59. Sydney Smith to Lady Holland, 1816, quoted in Parker, "Lady Davy in Her Letters," 82. Smith seems to be alluding to Goethe's novel *Elective Affinities* (1809), with its chemical metaphors for coupling and uncoupling, while discounting the possibility of an extramarital attachment in this case. His mention of Jane's disappointment with the "powers of Chemistry" suggests Davy lacked sexual potency, but that remark may simply reflect his jealousy at Davy's successful courting of a woman to whom he was himself attracted. The circulation of crude sexual gossip regarding the Davys' marriage is also attested by fragments of doggerel preserved in National Library of Scotland, MS 24601, f. 10.

60. Humphry Davy to John Davy, 30 October 1823, Davy Letters Project.

61. Record of conversation between Sir Charles Clarke, Charles Babbage, and Sir Benjamin Collins Brodie, 18 March 1839, in Miscellaneous Correspondence of Sir Benjamin Collins Brodie, University of Leicester Library, MS 212.

62. J. Davy, *Fragmentary Remains*, 142–43.

63. [Anon.,] "Sir Humphry Davy, Bart.," 69.

64. [Brewster,] "Memoirs of the Life of Sir Humphry Davy, Bart.," 118.

65. Humphry Davy to Jane Davy, 27 September 1828 and 20 January 1829, Davy Letters Project.

66. The prevalence of these stories may also reflect the strengthening of the ideal of companionate marriage in this period, for which see Stone, *Family, Sex and Marriage*, 217–53; Vickery, *Gentleman's Daughter*, 59–72; Hunt and Jacob, "Affective Revolution."

67. Crowther, *British Scientists*, 1:20–21, 65, 71, 79.

CHAPTER 4

1. Carlyle, "Signs of the Times," 66.

2. [Brewster,] "Memoirs of the Life of Sir Humphry Davy," 102.

3. Schaffer, "Scientific Discoveries." On this topic, see also Schaffer, "Making up Discovery"; Brannigan, *Social Basis of Scientific Discoveries*; Nickles, "Discovery."

4. Davy's linking of machinery with performance, the body, and organic

phenomena in general was symptomatic of the era. For a wide-ranging analysis, see Tresch, *Romantic Machine*.

5. For background on Galvani, Volta, and the electrical battery, see Pancaldi, *Volta*, 178–256; Kipnis, "Luigi Galvani"; Strickland, "Galvanic Disciplines"; Pera, *Ambiguous Frog*; Sudduth, "Voltaic Pile."

6. J. Davy, Memoirs of The Life of Sir Humphry Davy, 1:307.

7. Royal Institution MSS: HD/20/a, 20; HD/13/h, 11–12.

8. Humphry Davy to Davies Giddy, 3 July 1800, Davy Letters Project.

9. Humphry Davy to Davies Giddy, 20 October 1800, Davy Letters Project.

10. Humphry Davy to Samuel Taylor Coleridge, 26 November 1800, Davy Letters Project.

11. On Nicholson's role as a scientific journalist, see Watts, "'We Want No Authors.'"

12. H. Davy, "An Account of Some Galvanic Combinations."

13. H. Davy, "Letter to Mr. Nicholson."

14. H. Davy, "An Essay on Heat, Light and the Combinations of Light."

15. Robert Southey to Humphry Davy, 26 July 1800, Royal Institution MSS: HD/27/B/5.

16. H. Davy, "Outlines of a View of Galvanism."

17. Pancaldi, "On Hybrid Objects and Their Trajectories," 257–58.

18. Humphry Davy to James Watt, 7–8 January 1801, Davy Letters Project.

19. Humphry Davy to James Watt, 29 December 1801, Davy Letters Project.

20. Priestley, "Observations and Experiments Relating to the Pile of Volta."

21. H. Davy, "Bakerian Lecture: On Some New Phenomena," 33n. For other attempts at the time to revive versions of the phlogiston theory, see Sudduth, "Eighteenth-Century Identifications of Electricity with Phlogiston"; Boantza, "Phlogistic Role of Heat in Chemical Revolution"; Kim, "'Instrumental' Reality of Phlogiston."

22. Dalhousie University Archives, James Dinwiddie MSS: E.6, 14 January 1809; B.77, 15 January 1809.

23. Morus, "Radicals, Romantics and Electrical Showmen." See also Morus, "'More the Aspect of Magic than Anything Natural.'"

24. Dalhousie University Archives, James Dinwiddie MSS: B.75, 1 March 1808; B.77, 7 March 1809.

25. Aldini, *An Account of the Late Improvements in Galvanism*.

26. [Anon.,] "Account of Late Improvements in Galvanism." Walter Houghton, in *The Wellesley Index to Victorian Periodicals*, assigns the review to Davy, although Henry Brougham's manuscript index lists the author as John Thomson. Notwithstanding the note in *the Wellesley Index* (4:779), Thomson did have a documented interest in galvanism; see Golinski, *Science as Public Culture*, 207n., and the sources cited there. See also Fullmer, *Sir Humphry Davy's Published Works*, 46; and, for Davy's authentic comments on Aldini's work, H. Davy, "Observations relating to the Progress of Galvanism."

27. [Anon.,] "Account of Late Improvements in Galvanism," 195, 196.

28. H. Davy, "Outlines of a View of Galvanism."

29. Dalhousie University Archives, James Dinwiddie MSS: E.7 [June 1810].

30. Priestley, "Observations and Experiments Relating to the Pile of Volta."

31. Ritter's experiments were reported in: "G. M., Dr.," "On the State of Galvanism and Other Scientific Pursuits in Germany."

32. H. Davy, "Bakerian Lecture: On Some Chemical Agencies of Electricity."

33. [Brougham,] review of Davy's First Bakerian Lecture, 398.

34. H. Davy, "Bakerian Lecture: On Some New Phenomena."

35. H. Davy, "Bakerian Lecture for 1809."

36. H. Davy, "Lecture Introductory to Electro-Chemical Science," at 281–282.

37. [Brougham,] review of Davy's Second Bakerian Lecture, 395.

38. H. Davy, *Collected Works*, 5:106n.

39. John Davy quoted in ibid., 5:107n.

40. Sylvester, *Elementary Treatise on Chemistry*, 119–35.

41. E. Walker, *Philosophical Essays*.

42. Gibbes, *Phlogistic Theory*, 20.

43. Kurzer, "William Hasledine Pepys."

44. Sutton, "Politics of Science in Early Napoleonic France"; Mac Arthur, "Davy's Differences with Gay-Lussac and Thenard."

45. H. Davy, *Collected Works*, 5:284–311, 5:312–48.

46. H. Davy, *Collected Works*, 5:345.

47. Golinski, *Science as Public Culture*, 225–34.

48. Dalhousie University Archives, James Dinwiddie MSS: E.7, 2 February 1811; E.11, 16 March 1811.

49. J. Davy, "Account of an Experiment."

50. On the pedagogical tradition in chemistry, see Hannaway, *Chemists and the Word*; Moran, *Distilling Knowledge*.

51. Black's lectures were published after his death: Black, *Lectures on the Elements of Chemistry*.

52. Bensaude-Vincent, *Lavoisier*; Kim, *Affinity, That Elusive Dream*.

53. Bensaude-Vincent, "View of Chemical Revolution"; Bensaude-Vincent and Abbri, *Lavoisier in European Context*.

54. Compare the perspective of Bensaude-Vincent, *Lavoisier*, with earlier studies of the chemical revolution, including F. Holmes, " 'Revolution in Chemistry and Physics'"; Melhado, "Chemistry, Physics, and Chemical Revolution."

55. Nicholson's position is discussed in Golinski, "'Nicety of Experiment,'" 83–87.

56. H. Davy, *Elements of Chemical Philosophy*, viii.

57. [Thomson,] review of H. Davy, *Elements of Chemical Philosophy*, 372.

58. [Bostock,] review of H. Davy, *Elements of Chemical Philosophy*, 149.

59. J. Davy, *Memoirs of the Life of Sir Humphry Davy*, 1:437.

60. [Bostock,] review of H. Davy, *Elements of Chemical Philosophy*, 158.

61. On this tradition, see Christie, "Historiography of Chemistry."

62. H. Davy, *Elements of Chemical Philosophy*, 19.

63. Ibid., 20. Priestley had conferred recognition on Davy as a successor to his own style of chemical discovery in a letter written in 1801: Joseph Priestley to Humphry Davy, 31 October 1801, Royal Institution MSS: HD/9, 67–70.

64. H. Davy, *Elements of Chemical Philosophy*, 23.

65. Ibid., 37, 109.

66. [Bostock,] review of H. Davy, *Elements of Chemical Philosophy*, 155.

67. Humphry Davy to William Clayfield, 28 August 1812, Davy Letters Project. The Royal Institution Archives contain two copies of Davy's *Elements of Chemical Philosophy*, annotated with comments in his hand and Faraday's (Royal Institution MSS: HD/24/1 and HD/24/2). These were appar-

ently prepared with an eye to a second edition, and the first volume includes a handwritten draft of an advertisement for such an edition, dated 31 October 1813.

68. J. Davy, *Memoirs of the Life of Sir Humphry Davy*, 1:455.

69. H. Davy, *Elements of Chemical Philosophy*, 136.

70. [Bostock,] review of H. Davy, *Elements of Chemical Philosophy*, 152, 157.

71. [Anon.,] review of H. Davy, *Elements of Chemical Philosophy*, 75.

72. Murray, *System of Chemistry*, 1:vii.

73. See the "Advertisement," drafted for a proposed second edition, in H. Davy, *Collected Works*, 4:xi–xv.

74. Abbri, "Romanticism versus Enlightenment."

75. H. Davy, *Elements of Chemical Philosophy*, 274.

76. Ibid.

77. Ibid., 279. On Davy's interest in Boscovich, see Siegfried, "Boscovich and Davy."

78. Dalhousie University Archives, James Dinwiddie MSS: E.6, 11 March 1809.

79. Ure, "Experiments on Relation between Muriatic Acid and Chlorine," 332.

CHAPTER 5

1. Dalhousie University Archives, James Dinwiddie MSS: B.81; E.9; and E.11, January–April 1811. See also E.6, 11 March 1809.

2. Golinski, *Science as Public Culture*, 11–37, 59–60; Donovan, *Philosophical Chemistry*; Sumner, "Michael Combrune, Peter Shaw."

3. This argument is made effectively in Roberts, Schaffer, and Dear, *Mindful Hand*, xiii–xxvii, 1–8.

4. Bacon's framing of the issue is discussed in Gaukroger, "Biography as a Route to Understanding Early Modern Natural Philosophy."

5. On Watt, see MacLeod, "James Watt, Heroic Invention"; MacLeod, *Heroes of Invention*; Miller, "'Puffing Jamie'"; Miller, *James Watt, Chemist*; Miller, "Usefulness of Natural Philosophy."

6. Humphry Davy to James Watt Jr., 10 February 1821, Davy Letters Project.

7. Eulogy for James Watt, 18 June 1824, in H. Davy, *Collected Works*, 7:141–45, at 141.

8. Gregory Watt to Humphry Davy, 11 October 1800 and 7 February 1801, in Royal Institution MSS: HD/26/G/6 and HD/26/G/7.

9. Humphry Davy to Davies Giddy, 22 February 1799, Davy Letters Project.

10. Humphry Davy to Samuel Taylor Coleridge, 8/9 June 1800, Davy Letters Project; compare Royal Institution MSS: HD/13/c, 49.

11. See the draft of a chemical dialogue in Royal Institution MSS: HD/14/b, 1–61, where the character named "Philos[opher]" is corrected to "the Unknown."

12. Notes by John Davy in Royal Institution MSS: HD/21/d, 270.

13. Humphry Davy to John Tonkin, 12 January 1801, Davy Letters Project.

14. Humphry Davy to [John Tonkin?], (draft, 1801–02?) in Royal Institution MSS: HD/13/c, 39–43, at 42.

15. Compare: "Discourse Introductory to a Course of Lectures on Chemistry," 2:311–26, at 323, with the draft notes in Royal Institution MSS: HD/13/c, 48–65, at 56.

16. Royal Institution MSS: HD/13/c, 56, 52, 57.

17. H. Davy, *Collected Works*, 2:323.

18. Berman, *Social Change and Scientific Organization*, chap. 1.

19. Royal Institution Managers' Minutes, 29 June 1801, quoted in Berman, *Social Change and Scientific Organization*, 49. For Davy's work on tanning, see J. Davy, *Memoirs of the Life of Sir Humphry Davy*, 1:335–36; Paris, *Life of Sir Humphry Davy*, 110–14; Berman, *Social Change and Scientific Organization*, 49–54.

20. H. Davy, *Collected Works*, 2:317.

21. H. Davy, "Account of Some Experiments," 272.

22. Sandford, *Thomas Poole and His Friends*, 1:276; 2:102, 287.

23. H. Davy, *Consolations in Travel*, [vii].

24. Humphry Davy to Thomas Poole, 26 March 1828, Davy Letters Project; Thomas Poole to John Davy, 12 November 1835, Royal Institution MSS: HD/26/D/56.

25. Berman, *Social Change and Scientific Organization*, 65–68, at 64 (plate 10).

26. Bord, *Science and Whig Manners*, 106–18. See also Miller, "Mannered Science and Political Identity"; Janković, *Reading the Skies*, 131–37 (on the Georgic tradition).

27. H. Davy, *Collected Works*, 2:316. On Cullen, see Golinski, *Science as Public Culture*, 13–37; on Scottish improvement in general, see Jonsson, *Enlightenment's Frontier*.

28. H. Davy, *Collected Works*, 7:197.

29. See also H. Davy, "On the Analysis of Soils."

30. H. Davy, *Collected Works*, 7:198, 304–5.

31. [Anon.,] "Elements of Agricultural Chemistry," 280.

32. Miles, "Sir Humphrey Davie." See also B. Cohen, *Notes from the Ground*, 30–42, 81–85.

33. [Anon.,] "Elements of Agricultural Chemistry," 251.

34. Fullmer, "Humphry Davy and Gunpowder Manufactory."

35. Humphry Davy to J. G. Children, 14 July 1812, Davy Letters Project.

36. Humphry Davy to W. Parkes, 1 August 1812, Davy Letters Project. Only a fragment of the letter is available (quoted in Fullmer, "Humphry Davy and Gunpowder Manufactory," 173–74). The document is believed to be in private hands.

37. Fullmer, "Humphry Davy and Gunpowder Manufactory," 177–78.

38. Humphry Davy to J. G. Children, 19 June 1813, Davy Letters Project.

39. Humphry Davy to J. G. Children, 19 July 1813, Davy Letters Project. The letter is quoted in Fullmer, "Humphry Davy and Gunpowder Manufactory," 182–83; its date has been corrected by the editors of the Davy Letters Project.

40. Humphry Davy to J. G. Children, 22 July 1813, Davy Letters Project.

41. Humphry Davy to J. G. Children, 23 July 1813, Davy Letters Project.

42. Fullmer, "Humphry Davy and Gunpowder Manufactory," 174–79. The letter by Eleanor Anne Porden to Mary Flaxman, 11 April 1812, reflects public apprehension at the time of his marriage that Davy would be corrupted by the wealth he was inheriting. If so, Porden remarked, "the beginning of his rank will be the end of his greatness" (Derbyshire Record Office, Matlock, DD3311/25/1/17). It is also worth noting that Davy wrote in October 1813 to the Scottish engineer James Rennie, who had also been involved in the gunpowder project, that his motive for quitting was that "I was not in want of any addition of income" (Humphry Davy to James Rennie, 9 October 1813, Davy Letters Project).

43. [Mr.] Burton to J. G. Children, undated letter quoted in Fullmer, "Humphry Davy and Gunpowder Manufactory," 187–88.

44. [Mr.] Burton to Humphry Davy, 25 July 1813, quoted in Fullmer, "Humphry Davy and Gunpowder Manufactory," 187.

45. Fullmer, "Humphry Davy and Gunpowder Manufactory," 186, 192.

46. Humphry Davy to J. G. Children, 26 July 1813, Davy Letters Project.

47. James, "How Big Is a Hole?" 183.

48. Royal Institution MSS: HD/14/g, 27, 29, 30. For notes by Davy and Faraday critical of Stephenson's claim, see also Royal Institution MSS: HD/10, 4–15.

49. [Winch,] "On Safe-Lamps for Coal Mines," 460, quoted in James, "How Big Is a Hole?" 203.

50. James, "How Big Is a Hole?" 207.

51. In a letter of September 1816, Davy reminded Lambton of their earlier connection and urged him to renew his "efforts in the cause of liberal & independent politics" and his "attacks upon corruption" (Humphry Davy to J. G. Lambton, 9 September 1816, Davy Letters Project).

52. See the account of the event in the *Morning Post*, 18 October 1817 (no. 14584).

53. Humphry Davy to John Hodgson, 12 July 1816, Davy Letters Project, quoted in James, "How Big Is a Hole?" 207.

54. Humphry Davy to Earl of Strathmore, 10 November 1817, Davy Letters Project.

55. Humphry Davy to James Losh, 11 November 1817, Davy Letters Project. For Losh's friendships in Davy's circle, see Humphry Davy to James Losh, 28 December 1803, Davy Letters Project; *Oxford Dictionary of National Biography*, s.v. "Losh, James," accessed 2 April 2015, http://www.oxforddnb.com.

56. Earl of Strathmore to Davy, 17 November 1817, and James Losh to Davy, 13 November 1817, transcribed with Davy's letters to these individuals in the database of the Davy Letters Project.

57. Joseph Banks to Humphry Davy, 30 October 1815, in Royal Institution MSS: HD/9, 299–300.

58. Miller, "Between Hostile Camps," 23–30.

59. Humphry Davy to Thomas Poole, 10 December 1820, Davy Letters Project.

60. Miller, "Usefulness of Natural Philosophy."

61. James, "Davy in the Dockyard."

62. "Ships' Copper," *Times*, 10 January 1825 (no. 12546), 3.

63. Davy, "Further Researches on Preservation of Metals," 341.

64. [Anon.,] "Humbugs of the Age," 92.

65. *Times*, 16 October 1824 (no. 12472), 2, quoted in James, "Davy in the Dockyard," 218.

66. The petition for a naval appointment of Mr. Millet, apparently married to Davy's sister Betsy, was mentioned frequently in letters from Davy to his mother, Grace Davy, during 1824–25 (all included in Davy Letters Project). On 21 April 1825, Davy wrote that Millet had gained an appointment, though perhaps not a permanent one. On 10 December 1825, he mentioned that Millet was to be tried for causing a death; Davy hoped for an acquittal but felt implicated by having recommended the man for an appointment. On 5 January 1826, Davy recorded that the death was due to accidental discharge of a pistol. On 25 February 1826, he concluded—understandably—that Millet lacked the scientific talents to rise in the navy.

67. Humphry Davy to J. G. Children, 29 October 1824, Davy Letters Project (emphasis in the original).

68. Paris, *Life of Sir Humphry Davy*, 399.

69. J. Davy, *Memoirs of the Life of Sir Humphry Davy*, 2:186.

70. *Chemist* 2:31 (1824–25), 47. On this journal, see Brock, "London Chemical Society of 1824"; Russell, *Science and Social Change*, 139–46.

71. *Chemist* 1 (1824), vii.

72. *Chemist* 1:16 (1824), 252.

73. Humphry Davy to J. G. Children, [late October 1824], Davy Letters Project.

74. Humphry Davy to Jane Davy, 25 April 1827, Davy Letters Project.

75. Humphry Davy to Jane Davy, 25 September 1827 and 1 June 1828, Davy Letters Project.

76. H. Davy, *Consolations in Travel*, 224.

77. Ibid., 255.

CHAPTER 6

1. Humphry Davy's journal of the 1806 tour is in Royal Institution MSS: HD/15/b; with extracts in H. Davy, *Collected Works*, 7:146–68.

2. This tour is described in Humphry Davy's letters to John Davy, 18 March 1814; to William Sotheby, 12 April 1814; and to Robert Liston, 15 February 1815, all in Davy Letters Project. See also Faraday's journal of the trip, published as Faraday, *Curiosity Perfectly Satisfyed*.

3. Sources for this journey include Humphry Davy's letters to W. T. Brande, 20 May 1820; and to Grace Davy, 16 June 1820, Davy Letters Project.

4. Sources for this journey include Humphry Davy's letters to Jane Davy, 15 March 1827; and to Davies Giddy, 18 March 1827, Davy Letters Project.

5. Tobin, *Journal of a Tour*.

6. Humphry Davy to Robert Liston, 15 February 1815, Davy Letters Project.

7. See Faraday, *Curiosity Perfectly Satisfyed*, 23–30, 73–77; H. Davy, "Some Experiments and Observations on Colours."

8. H. Davy, "Some Observations and Experiments on Papyri."

9. On the relations between travel and the methods of the fieldwork sciences, see Bourguet, Licoppe, and Sibum, *Instruments, Travel and Science*; Bourguet, "Explorer"; Craciun, "What Is an Explorer?"; Agar and Smith, *Making Space for* Science; Finnegan, "Spatial Turn"; Powell, "Geographies of Science"; Secord, "Knowledge in Transit."

10. Humphry Davy to William Sotheby, 26 March 1820, Davy Letters Project.

11. Humphry Davy to John Davy, 18 March 1814, Davy Letters Project.

12. Rudwick, "Emergence of Visual Language."

13. Examples of landscapes and sketches of exposed strata include those in Royal Institution MSS: HD/15/a. Compare the sketch map of the horizontal distribution of strata in Royal Institution MSS: HD/15/c, 178.

14. On the aesthetic dimension of travel in this period, see Adler, "Travel as Performed Art"; Cardinal, "Romantic Travel"; Elsner and Rubiés, *Voyages and Visions*; Gilroy, *Romantic Geographies*; Leask, *Curiosity and Aesthetics of Travel Writing*; Thompson, *Suffering Traveller*.

15. For instances, see Humphry Davy's letters to Jane Davy, 2 June 1827, 10 August 1827, and 25 September 1827; Humphry Davy to Thomas Andrew Knight, 30 July 1827, all in Davy Letters Project.

16. Discussions of Davy's *Consolations* include Secord, *Visions of Science*, 24–51, and Abbri, "Romanticism versus Enlightenment." It is fair to say that the work remains something of an enigma, notwithstanding these attempts at contextualization. Parallels have been drawn with dialogues by later Romantic-era natural philosophers that were also published posthumously: H. C. Ørsted's *The Soul in Nature* (1852) and F. W. J. Schelling's *Clara* (1861). June Fullmer suggested that one influence on Davy's conception of the book was "The Vision" in volume 4 of Abraham Tucker's *The Light of Nature Pursued* (1768), which Davy read in his Bristol days. See Fullmer, *Young Humphry Davy*, 115.

17. Humphry Davy to J. J. Berzelius, 19 October 1813, Davy Letters Project.

18. H. Davy, *Humphry Davy on Geology*, 116–29, and, for identification of Greenough, 159n10. See also Siegfried and Dott, "Humphry Davy as Geologist."

19. H. Davy, *Humphry Davy on Geology*, 121, 123.

20. Ibid., 127.

21. Ibid., 128.

22. Royal Institution MSS: HD/12, 64.

23. Royal Institution MSS: HD/14/g, 79.

24. Rossi, *Dark Abyss of Time*; Rudwick, *Bursting Limits of Time*.

25. On Hutton, see Rudwick, *Bursting Limits of Time*, 158–72; Dean, *James Hutton and History of Geology*.

26. H. Davy, *Humphry Davy on Geology*, 45.

27. Ibid., 54.

28. On the decline of Darwin's reputation, see Garfinkle, "Science and Religion in England."

29. Royal Institution MSS: HD/12, 63–67.

30. Royal Institution MSS: HD/12, 63.

31. Royal Institution MSS: HD/12, 66.

32. Lyell, *Principles of Geology* (1997), xxxi–xxxii, [2] (for the illustration).

33. Royal Institution MSS: HD/14/g, 60–84 (written from the back of the book toward the front, quotation on 77–76).

34. On Buckland's work, see: Rudwick, *Worlds before Adam*, 73–87.

35. H. Davy, *Six Discourses Delivered Before the Royal Society*, 51–52.

36. Humphry Davy to Joseph Cottle, 11 June 1823, Davy Letters Project.

37. Ambrosio was generally identified with Monsignor Lavinio de' Medici Spada, the papal vice-legate in Ravenna, in whose house Davy had lodged for a couple of months in 1827. (See the comments by John Davy in *Memoirs of the Life of Sir Humphry Davy*, 2:374–76.) On the same authority, Eubathes in Dialogue IV of *Consolations* was taken to be a portrait of Davy's friend William Hyde Wollaston. Anne Treneer, in *The Mercurial Chemist*, 230, made the suggestion that Onuphrio was based on William Lamb, later Lord Melbourne, but this does not seem likely. The *Oxford Dictionary of National Biography* says that Lamb studied briefly in Glasgow but was uninfluenced by Scottish Enlightenment thinking and uninterested in religious debate. "Onuphrio" is the Italian equivalent of "Humphry," and in that respect this character might be taken as another stand-in for the author.

38. H. Davy, *Consolations in Travel*, 136–37.

39. Ibid., 149.

40. Ibid., 150.

41. Ibid., 151.

42. Boethius, *Consolation of Philosophy*; Lucretius, *On the Nature of Things*.

43. H. Davy, *Consolations in Travel*, 281–82.

44. Darwin, *Origin of Species*, 460.

45. Tobin, *Journal of a Tour*, 96 (June 1828).

46. Ibid., 120–21.

47. Humphry Davy to Jane Davy, 14 July 1828, Davy Letters Project.

48. Humphry Davy to Jane Davy, 29 July 1828, Davy Letters Project.

49. Humphry Davy to Jane Davy, 3 December 1828, Davy Letters Project.

50. Tobin, *Journal of a Tour*, 210 (24 February 1829).

51. Humphry Davy to John Davy, 23 February 1829, Davy Letters Project.

52. Humphry Davy to Jane Davy, 1 March 1829, Davy Letters Project.

53. Royal Institution MSS: HD/14/i, 55–56.

54. H. Davy, *Consolations in Travel*, [v].

55. [Anon.,] "Consolations in Travel," *Dublin Literary Gazette*, 197.

56. [Anon.,] "Sir H. Davy's Posthumous Work," 82.

57. [Anon.,] "Consolations in Travel," *Dublin Literary Gazette*, 199.

58. [Anon.,] "Sir Humphry Davy's Consolations in Travel," 392.

59. See, for example, [Anon.,] "Sir Humphry Davy, Bart.," 39–85; W. Walker, *Memoirs of the Distinguished Men of Science of Great Britain*, 65–71.

60. The London editions were 1830, 1831, 1833, 1838, 1851 (with illustrations), 1853, 1869, 1889, and 1895. Also 1840 (with *Salmonia*), 1839–40 (in H. Davy, *Collected Works*). The American editions were 1830 (Philadelphia), 1870 (Boston). Reported translations include 1833 (German), 1834 (Swedish), 1836 (Dutch), 1869 (French), 1878 (Spanish).

61. H. Davy, *Consolations in Travel*, 67.

62. Ibid., 71.

63. The poem is in Royal Institution MSS: HD/14/e, 128–30. See also: HD/14/k, 72, 94; and HD/14/m, 114–15, for phrases in German and Slovenian apparently written by "Pepina" in Davy's notebook.

64. H. Davy, *Collected Works*, 1:433–36.

65. Ibid., 1:436–37.

66. Ibid., 1:430.

67. [Anon.,] "Sir Humphry Davy's Consolations in Travel," 398.

68. H. Davy, *Collected Works*, 1:429, 438.

69. Ibid., 1:438.

70. H. Davy, *Collected Works*, 1:332–333n, 430–31. On Monsignor Spada, see also Davy's letters to Thomas Andrew Knight, 1 April 1827; to John Davy, 21 December 1828; and to Jane Davy, 1 March 1829, all in Davy Letters Project.

71. H. Davy, *Consolations in Travel*, 160–62.

72. [Anon.,] "Sir H. Davy's Posthumous Work," 83.

73. Royal Institution MSS: HD/13/b, 12, 14.

74. Royal Institution MSS: HD/13/b, 22–23, 31.

75. H. Davy, *Consolations in Travel*, 206–7.

76. Ibid., 213.

77. Ibid., 219.

78. On the dispute between the London surgeons John Abernethy and William Lawrence, see: Jacyna, "Immanence or Transcendence"; Levere, *Poetry Realized in Nature*, 45–52.

79. [Anon.,] "Consolations in Travel," *Anti-Infidel*, 174. Compare [Anon.,]

"Sir Humphry Davy's Consolations in Travel," 405, and [Anon.,] "Consolations in Travel," *British Magazine*.

80. Royal Institution MSS: HD/14/1, 31–32.

81. H. Davy, *Consolations in Travel*, 191.

82. Ibid., 191–92.

83. The expanded comments are in H. Davy, *Salmonia* (2nd ed., 1829), 73–74. See also Davy's notes for the revised edition of the book: Royal Institution MSS: HD/14/m, 147–60.

84. H. Davy, *Consolations in Travel*, 281.

85. Tobin, *Journal of a Tour*, 242.

86. Humphry Davy to Jane Davy, 21 August 1828, Davy Letters Project.

87. Royal Institution MSS: HD/22/c, 39.

EPILOGUE

1. H. Davy, *Collected Works*, 1:445.

2. Humphry Davy to Jane Davy, 19 May 1827, Davy Letters Project.

3. See, for example, Humphry Davy to Jane Davy, 15 March 1827, and 2 June 1827, Davy Letters Project.

4. H. Davy, *Collected Works*, 1:446.

5. Humphry Davy to Davies Giddy, 18 March 1827, Davy Letters Project.

6. Davy, "Account of Some Experiments on Torpedo."

7. Testifying to Davy's bodily sensitivity when he began this inquiry in Genoa in March 1814, Michael Faraday recorded that the fish sometimes gave such weak shocks "that I could not feel them but Sir H. Davy did." Faraday, *Curiosity Perfectly Satisfyed*, 67.

8. Hazlitt continued: "I am not a bit the wiser for any of his discoveries, nor I never met with any one that was." Hazlitt, "Indian Jugglers," 123. He seems to have spoken for many of his contemporaries in discounting Davy's stature as a literary figure. Compare George Eliot's well-known scene, in *Middlemarch* (1871–72), in which Mr. Brooke reminds his dinner guests that Davy was a poet as well as an agricultural chemist. By the second half of the nineteenth century—or even by the 1830s, when the scene was set—his poetic ambitions had been largely forgotten. Eliot, *Middlemarch*, 38.

9. Morrell and Thackray, *Gentlemen of Science*; Snyder, *Philosophical Breakfast Club*.

10. Quoted in Fullmer, "Davy's Sketches of His Contemporaries," 134.

11. Quoted in Smiles, *Publisher and His Friends*, 2:208. See "UK Reading Experience Database, Record No. 27339," accessed 11 August 2014, http://www.open.ac.uk/Arts/reading/UK/record_details.php?id=27339.

12. Editorial [by Thomas Hodgskin], *Chemist* 2:33, 78.

13. Morrell and Thackray, *Gentlemen of Science*, 148–57.

14. Brande, *Manual of Chemistry*.

15. Secord, *Visions of Science*; Fyfe and Lightman, *Science in the Marketplace*.

16. Tresch, *Romantic Machine*.

17. Cunningham and Jardine, *Romanticism and the Sciences*; Levere, *Poetry Realized in Nature*.

18. Humboldt, *Views of Nature*, 154. Studies of Humboldt's mode of scientific travel include Dettelbach, "Humboldtian Science"; Nicholson, "Alexander von Humboldt"; Pratt, *Imperial Eyes*, 111–43; Cannon, *Science in Culture*, 73–110; Walls, *Passage to Cosmos*.

19. Lyell, *Principles of Geology* (1830–33), 1:145; 2:117–18, 271. At another point, he used the label in a more positive manner, calling Davy "a late distinguished philosopher" (2:271).

20. Darwin, Notebook N, 36, in Darwin Online. The source is H. Davy, *Consolations in Travel*, 211.

21. Smiles, *Self-Help*, 14, 78–79, 290–91.

22. Crowther, *British Scientists*, 1:15.

23. H. Davy, *Consolations in Travel*, 189.

Bibliography

MANUSCRIPT AND ONLINE SOURCES

Dalhousie University Archives, Halifax, Nova Scotia. Manuscripts of James Dinwiddie (MS-2-726. Acc.# 3-99).

Darwin Online. Charles Darwin, Notebook N (Metaphysics and Expression, 1838–39). Accessed 2 April 2015. http://darwin-online.org.uk.

Davy Letters Project. Letters of Humphry Davy and His Circle. http://www.davy-letters.org.uk. To be published as *The Collected Letters of Sir Humphry Davy* (Tim Fulford and Sharon Ruston, general editors; with Jan Golinski, Frank James, and David Knight, advisory editors). 4 vols. Oxford: Oxford University Press, 2017.

Derbyshire Record Office, Matlock, Derbyshire. Letter from Eleanor Anne Porden to Mary Ann Flaxman, 11 April 1812 (DD3311/25/1/17).

Huntington Library, San Marino, California. Letters of Robert Southey to William Taylor, 1799–1813 (MSS HM 4819, 4820, 4821, 4823, 4825, 4826, 4827, 4832, 4847, 4870).

National Library of Scotland, Edinburgh. Letter by Leonard Horner to Alexander Marcet, 10 April 1821 (MS 9818, f.88–89). Doggerel poetry on Sir Humphry Davy's marriage (MS 24601, f. 10).

Royal Institution of Great Britain, London. Manuscripts of Sir Humphry Davy, MSS HD/1-HD/29.

Special Collections, University of Leicester Library, Leicester. Miscellaneous correspondence of Sir Benjamin Collins Brodie, c. 1855–1862 (MS 212).

BOOKS, JOURNAL ARTICLES, AND DISSERTATIONS

Abbri, Ferdinando. "Romanticism versus Enlightenment: Sir Humphry Davy's Idea of Chemical Philosophy." In *Romanticism in Science: Science in Europe, 1790–1840*, edited by Stefano Poggi and Maurizio Bossi, 31–45. Dordrecht: Kluwer, 1993.

Adams, James Eli. "The Hero as Spectacle: Carlyle and the Persistence of Dandyism." In *Victorian Literature and the Victorian Visual Imagination*, edited by Carol T. Christ and John O. Jordan, 213–32. Berkeley: University of California Press, 1995.

Adler, Judith. "Travel as Performed Art." *American Journal of Sociology* 94 (1989): 1366–91.

Agar, Jon, and Crosbie Smith, eds. *Making Space for Science: Territorial Themes in the Shaping of Knowledge*. Basingstoke, UK: Macmillan, 1998.

Aldini, Giovanni. *An Account of the Late Improvements in Galvanism: with a Series of Curious and Interesting Experiments . . . To Which Is Added an Appendix Containing Experiments on the Body of a Malefactor Executed at Newgate*. London: Cuthell and Martin and J. Murray, 1803.

Altick, Richard D. *The Shows of London*. Cambridge, MA: Harvard University Press, 1978.

[Anon.] "An Account of the Late Improvements in Galvanism, with a Series of Curious and Interesting Experiments." *Edinburgh Review* 3, no. 5 (October 1803): 194–98.

[———.] Review of Davy, *Elements of Chemical Philosophy*. *Quarterly Review* 8 (September 1812): 65–86.

[———.] "Sir Humphrey Davy and His Visit to Paris." *Examiner* 304 (24 October 1813): 673–75.

[———.] "Elements of Agricultural Chemistry in a Course of Lectures for the Board of Agriculture by Sir Humphry Davy." *Edinburgh Review* 22, no. 44 (January 1814): 251–81.

[———.] "The Humbugs of the Age. No. III—Sir Humphrey Davy." *John Bull Magazine and Literary Recorder* 1 (1824): 89–92.

[———.] "Sir Humphry Davy, Bart., LL.D., F.R.S., M.R.L.A." *Annual Biography and Obituary* 14 (1830): 39–85. London: Longman, Rees, Orme, Brown, and Green.

[————.] "Sir H. Davy's Posthumous Work." *Athenaeum* 120 (13 February 1830): 82–83.

[————.] "Sir Humphry Davy's Consolations in Travel." *Monthly Review*, 3rd ser., 13, no. 55 (March 1830): 391–408.

[————.] "Consolations in Travel, or the Last Days of a Philosopher, by Sir Humphry Davy, Bart." *Dublin Literary Gazette* 13 (27 March 1830): 197–99.

[————.] "Sir Humphry Davy." *Mirror of Literature, Amusement, and Instruction* 472 (22 January 1831): 63.

[————.] "The Life of Sir Humphry Davy, Bart." *Monthly Review*, 4th ser., 1, no. 3 (March 1831): 364–85.

[————.] "Consolations in Travel: Or the Last Days of a Philosopher." *Anti-Infidel* 1, no. 11 (12 August 1831): 174.

[————.] "Consolations in Travel: Or the Last Days of a Philosopher." *British Magazine* 1 (March 1832): 38–40.

Babbage, Charles. *Reflections on the Decline of Science in England, and on Some of Its Causes*. London: B. Fellowes, 1830.

Baer, Marc. *Theatre and Disorder in Late Georgian London*. Oxford: Clarendon Press, 1992.

Bahar, Saba. "Jane Marcet and the Limits to Public Science." *British Journal for the History of Science* 34 (2001): 29–49.

Barry, Jonathan. "Piety and the Patient: Medicine and Religion in Eighteenth Century Bristol." In *Patients and Practitioners: Lay Perceptions of Medicine in Pre-Industrial Society*, edited by Roy Porter, 145–76. Cambridge: Cambridge University Press, 2003.

Bazerman, Charles. *Shaping Written Knowledge: The Genre and Activity of the Experimental Article in Science*. Madison: University of Wisconsin Press, 1988.

Beddoes, Thomas. *Notice of Some Observations Made at the Medical Pneumatic Institution*. Bristol: Biggs and Cottle, 1799.

Belcher, William. *Intellectual Electricity, Novum Organum of Vision, and Grand Mystic Secret*. London: Lee and Hurst, 1798.

Bell, Charles. *Letters of Sir Charles Bell, K.H., F.R.S.L. & E.* London: John Murray, 1870.

Ben-David, Joseph. *The Scientist's Role in Society: A Comparative Study*. 2nd ed. Chicago: University of Chicago Press, 1984.

Bensaude-Vincent, Bernadette. *Lavoisier: Mémoires d'une révolution*. Paris: Flammarion, 1993.

———. "A View of the Chemical Revolution through Contemporary Textbooks: Lavoisier, Fourcroy and Chaptal." *British Journal for the History of Science* 23 (1990): 435–60.

Bensaude-Vincent, Bernadette, and Ferdinando Abbri, eds. *Lavoisier in European Context: Negotiating a New Language for Chemistry*. Sagamore Beach, MA: Science History Publications USA, 1995.

Berman, Morris. *Social Change and Scientific Organization: The Royal Institution, 1799–1844*. Ithaca, NY: Cornell University Press, 1978.

Bermingham, Ann. "Urbanity and the Spectacle of Art." In *Romantic Metropolis: The Urban Scene of British Culture*, edited by James Chandler and Kevin Gilmartin, 151–76. Cambridge: Cambridge University Press, 2005.

Bertucci, Paola. "The Electrical Body of Knowledge: Medical Electricity and Experimental Philosophy in the Mid-Eighteenth Century." In *Electric Bodies: Episodes in the History of Medical Electricity*, edited by Giuliano Pancaldi and Paola Bertucci, 43–68. Bologna: CIS, 2001.

———. "Revealing Sparks: John Wesley and the Religious Utility of Electrical Healing." *British Journal for the History of Science* 39 (2006): 341–62.

Biagioli, Mario. *Galileo Courtier: The Practice of Science in the Age of Absolutism*. Chicago: University of Chicago Press, 1993.

Biagioli, Mario, and Peter Galison, eds. *Scientific Authorship: Credit and Intellectual Property in Science*. New York: Routledge, 2003.

Black, Joseph. *Lectures on the Elements of Chemistry, Delivered in the University of Edinburgh*. Edited by John Robison. 2 vols. Edinburgh: Mundell and Sons, 1803.

Boantza, Victor. "The Phlogistic Role of Heat in the Chemical Revolution and the Origins of Kirwan's 'Ingenious Modifications . . . Into the Theory of Phlogiston.'" *Annals of Science* 65 (2008): 309–38.

Boethius. *The Consolation of Philosophy*. Translated and with introduction and notes by P. G. Walsh. Oxford: Oxford University Press, 2000.

Bord, Joe. *Science and Whig Manners: Science and Political Style in Britain, c. 1790–1850*. Basingstoke, UK: Palgrave Macmillan, 2009.

[Bostock, John.] Review of Davy, *Elements of Chemical Philosophy*. *Monthly Review* 72 (October 1813): 148–58.

Bourguet, Marie-Noëlle. "The Explorer." In *Enlightenment Portraits*, edited by Michel Vovelle, translated by Lydia G. Cochrane, 257–315. Chicago: University of Chicago Press, 1999.

Bourguet, Marie-Noëlle, Christian Licoppe, and H. Otto Sibum, eds. *Instruments, Travel and Science: Itineraries of Precision from the Seventeenth to the Twentieth Century*. London: Routledge, 2003.

Brande, William Thomas. *A Manual of Chemistry*. London: John Murray, 1819.

Brannigan, Augustine. *The Social Basis of Scientific Discoveries*. Cambridge: Cambridge University Press, 1981.

Brewer, John. "Sensibility and the Urban Panorama." *Huntington Library Quarterly* 70 (2007): 229–49.

[Brewster, David.] "Memoirs of the Life of Sir Humphry Davy, Bart., L.L.D., F.R.S., Foreign Associate of the Institute of France." *Edinburgh Review* 63, no. 127 (April 1836): 101–35.

Brock, W. H. "The London Chemical Society of 1824." *Ambix* 14 (1967): 133–39.

[Brougham, Henry.] Review of Davy's First Bakerian Lecture. *Edinburgh Review* 11, no. 22 (January 1808): 390–98.

[———.] Review of Davy's Second Bakerian Lecture. *Edinburgh Review* 12, no. 24 (July 1808): 394–401.

———. "Davy." In *Lives of the Philosophers of the Time of George III*, by Henry Brougham. Vol. 1 of *The Works of Henry Lord Brougham*, 107–22. Edinburgh: Adam and Charles Black, 1872.

Browne, Janet. *Charles Darwin*. Vol. 1, *Voyaging*. Vol. 2, *The Power of Place*. New York: Alfred A. Knopf, 1995–2002.

Brush, Stephen G. *The History of Modern Science: A Guide to the Second Scientific Revolution, 1800–1950*. Ames: Iowa State University Press, 1988.

Bury, Charlotte. *The Diary of a Lady-in-Waiting, being the Diary Illustrative of the Times of George the Fourth, Interspersed with Original Letters from the Late Queen Caroline and from Other Distinguished Persons*. Edited by Francis Steuart. 2 vols. London: John Lane, the Bodley Head, 1908.

Cannon, Susan Faye. *Science in Culture: The Early Victorian Period*. New York: Dawson, 1978.

Cantor, Geoffrey. *Michael Faraday: Sandemanian and Scientist. A Study of Sci-*

ence and Religion in the Nineteenth Century. Basingstoke, UK: Macmillan, 1991.

———. "The Scientist as Hero: Public Images of Michael Faraday." In *Telling Lives in Science: Essays on Scientific Biography*, edited by Michael Shortland and Richard Yeo, 171–93. Cambridge: Cambridge University Press, 1996.

Cardinal, Roger. "Romantic Travel." In *Rewriting the Self: Histories from the Renaissance to the Present*, edited by Roy Porter, 135–55. London: Routledge, 1997.

Carlyle, Thomas. "Signs of the Times." In *Selected Writings*, edited by Alan Shelston, 61–85. Harmondsworth, UK: Penguin Books, 1971. Originally published 1829.

———. *Sartor Resartus. On Heroes, Hero-Worship, and the Heroic in History*. London: J. M. Dent, n.d. Originally published 1833–34.

Carter, Philip. "Tears and the Man." In Knott and Taylor, *Women, Gender and Enlightenment*, 156–73.

Cartwright, F. F. *The English Pioneers of Anaesthesia (Beddoes, Davy, Hickman)*. Bristol, UK: John Wright and Sons, 1952.

Chandler, James. "Introduction: Doctrines, Disciplines, Discourses, Departments." *Critical Inquiry* 35 (2009): 729–46.

Chilton, Donovan, and Noel G. Coley. "The Laboratories of the Royal Institution in the Nineteenth Century." *Ambix* 27 (1980): 173–203.

Christie, John R. R. "Historiography of Chemistry in the Eighteenth Century: Hermann Boerhaave and William Cullen." *Ambix* 41 (1994): 4–19.

Cohen, Benjamin R. *Notes from the Ground: Science, Soil, and Society in the American Countryside*. New Haven, CT: Yale University Press, 2009.

Cohen, Michèle. *Fashioning Masculinity: National Identity and Language in the Eighteenth Century*. London: Routledge, 1996.

———. "'Manners' Make the Man: Politeness, Chivalry, and the Construction of Masculinity, 1750–1830." *Journal of British Studies* 44 (2005): 312–29.

Colás, Santiago. "Aesthetics vs. Anaesthetic: How Laughing Gas Got Serious." *Science as Culture* 7 (1998): 335–53.

Coleridge, Samuel Taylor. *Collected Letters of Samuel Taylor Coleridge*. Edited by Earl Leslie Griggs. 6 vols. Oxford: Clarendon Press, 1956–71.

Combe, George. *Letter from George Combe to Francis Jeffrey, Esq. in Answer to His Criticism on Phrenology*. Edinburgh: John Anderson Jr., 1826.

Cooter, Roger, and Stephen Pumfrey. "Separate Spheres and Public Places: Reflections on the History of Science Popularisation and Science in Popular Culture." *History of Science* 22 (1994): 237–67.

Corneanu, Sorana. *Regimens of the Mind: Boyle, Locke, and the Early Modern Cultura Animi Tradition*. Chicago: University of Chicago Press, 2011.

Cottle, Joseph. *Reminiscences of Samuel Taylor Coleridge and Robert Southey*. New York: Wiley and Putnam, 1848.

Craciun, Adriana. "What Is an Explorer?" *Eighteenth-Century Studies* 45 (2011): 29–51.

Crompton, Louis. *Byron and Greek Love: Homophobia in 19th-Century England*. Berkeley: University of California Press, 1985.

Crowther, J. G. *British Scientists of the Nineteenth Century*. 2 vols. Harmondsworth, UK: Penguin Books, 1940.

Cunningham, Andrew, and Nicholas Jardine, eds. *Romanticism and the Sciences*. Cambridge: Cambridge University Press, 1990.

Cunningham, Andrew, and Perry Williams. "De-Centring the 'Big Picture': The Origins of Modern Science and the Modern Origins of Science." *British Journal for the History of Science* 26 (1993): 407–32.

Darnton, Robert. *Mesmerism and the End of the Enlightenment in France*. Cambridge, MA: Harvard University Press, 1968.

Darwin, Charles. *The Origin of Species by Means of Natural Selection; Or the Preservation of Favoured Races in the Struggle for Life*. Edited by J. W. Burrow. Harmondsworth, UK: Penguin Books, 1968. Originally published 1859.

Daston, Lorraine. "The Naturalized Female Intellect." *Science in Context* 5 (1992): 209–35.

Daston, Lorraine, and H. Otto Sibum. "Introduction: Scientific Personae and Their Histories." *Science in Context* 16 (2003): 1–8.

Davies, Gordon L. Herries. *Whatever Is Under the Earth: The Geological Society of London, 1807–2007*. Bath, UK: Geological Society of London, 2007.

Davy, Humphry. "An Essay on Heat, Light and the Combinations of Light." In *Contributions to Physical and Medical Knowledge, Principally from the*

West of England, edited by Thomas Beddoes, 4–147. Bristol, UK: Biggs and Cottle, 1799.

———. *Researches Chemical and Philosophical, Chiefly Concerning Nitrous Oxide, or Dephlogisticated Nitrous Air, and Its Respiration.* London: J. Johnson, 1800.

———. "Letter to Mr. Nicholson, Containing Notices Concerning Galvanism." *[Nicholson's] Journal of Natural Philosophy, Chemistry and the Arts,* 1st ser., 4 (February 1801): 527. Reprinted in H. Davy, *Collected Works,* 2:181–82.

———. "An Account of Some Galvanic Combinations, Formed by the Arrangement of Single Metallic Plates and Fluids, Analogous to the New Galvanic Apparatus of Mr. Volta." *Philosophical Magazine* 11 (1801–02): 202–6.

———. "Outlines of a View of Galvanism, Chiefly Extracted from a Course of Lectures on the Galvanic Phænomena, Read at the Theatre of the Royal Institution." *Journal of the Royal Institution* 1 (1802): 49–66. Reprinted in H. Davy, *Collected Works,* 2:188–209.

———. "Observations Relating to the Progress of Galvanism." *Journal of the Royal Institution* 1 (1802): 284–90. Reprinted in H. Davy, *Collected Works,* 2:221–28.

———. "A Discourse Introductory to a Course of Lectures on Chemistry" [26 April 1802]. Reprinted in H. Davy, *Collected Works,* 2:311–26.

———. "An Account of Some Experiments and Observations on the Constituent Parts of Certain Astringent Vegetables; And on Their Operation in Tanning." *Philosophical Transactions of the Royal Society* 93 (1803): 233–73.

———. "On the Analysis of Soils." *[Nicholson's] Journal of Natural Philosophy, Chemistry and the Arts,* 2nd ser., 12 (1805): 81–97.

———. "The Bakerian Lecture: On Some Chemical Agencies of Electricity." *Philosophical Transactions of the Royal Society* 97 (1807): 1–56.

———. "The Bakerian Lecture: On Some New Phenomena of Chemical Changes Produced by Electricity, Particularly the Decomposition of the Fixed Alkalies [*sic*], and the Exhibition of the New Substances Which Constitute Their Bases; And on the General Nature of Alkaline Bodies." *Philosophical Transactions of the Royal Society* 98 (1808): 1–44.

————. "Lecture Introductory to Electro-Chemical Science" [12 March 1808]. In H. Davy, *Collected Works*, 8:274–86.

————. "The Bakerian Lecture for 1809. On Some New Electrochemical Researches, on Various Objects, Particularly the Metallic Bodies, from the Alkalies [*sic*], and Earths, and on Some Combinations of Hydrogene." *Philosophical Transactions of the Royal Society* 100 (1810): 16–74.

————. *Elements of Chemical Philosophy*. Vol. 1, Part 1. London: J. Johnson, 1812.

————. "Some Experiments and Observations on the Colours Used in Painting by the Ancients." *Philosophical Transactions of the Royal Society* 105 (1815): 97–124.

————. "Some Observations and Experiments on the Papyri Found in the Ruins of Herculaneum." *Philosophical Transactions of the Royal Society* 111 (1821): 191–208.

————. "Further Researches on the Preservation of Metals by Electro-Chemical Means." *Philosophical Transactions of the Royal Society* 115 (1825): 328–46.

————. *Six Discourses Delivered Before the Royal Society, At Their Anniversary Meetings, on the Award of the Royal and Copley Medals*. London: John Murray, 1827.

————. *Salmonia: Or Days of Fly-Fishing; In a Series of Conversations; With Some Account of the Habits of Fishes Belonging to the Genus Salmo*. London: John Murray, 1828. 2nd ed., London: John Murray, 1829.

————. "An Account of Some Experiments on the Torpedo." *Philosophical Transactions of the Royal Society* 119 (1829): 15–18.

————. *Consolations in Travel; or, The Last Days of a Philosopher*. Edited by John Davy. London: John Murray, 1830.

————. *The Collected Works of Sir Humphry Davy*. Edited by John Davy. 9 vols. London: Smith, Elder, 1839–40.

————. *Humphry Davy on Geology: The 1805 Lectures for the General Audience*. Edited by Robert Siegfried and Robert H. Dott. Madison: University of Wisconsin Press, 1980.

Davy, John. "An Account of an Experiment Made in the College Laboratory, Edinburgh." *[Nicholson's] Journal of Natural Philosophy, Chemistry and the Arts*, 2nd ser. , 34 (1813): 68–72.

———. *Memoirs of the Life of Sir Humphry Davy*. 2 vols. London: Longman, 1836.

———. *Fragmentary Remains, Literary and Scientific, of Sir H. Davy*. London: Churchill, 1858.

Dean, Denis R. *James Hutton and the History of Geology*. Ithaca, NY: Cornell University Press, 1992.

Delbourgo, James. "Common Sense, Useful Knowledge, and Matters of Fact in the Late Enlightenment: The Transatlantic Career of Perkins's Tractors." *William and Mary Quarterly*, 3rd ser., 61 (2004): 643–84.

Desmond, Adrian. "Redefining the X Axis: 'Professionals,' 'Amateurs' and the Making of Mid-Victorian Biology: A Progress Report." *Journal of the History of Biology* 34 (2001): 3–50.

Desmond, Adrian, and James R. Moore. *Darwin*. New York: Warner, 1992.

Dettelbach, Michael. "Humboldtian Science." In *Cultures of Natural History*, edited by N. Jardine, J. A. Secord, and E. C. Spary, 287–304. Cambridge: Cambridge University Press, 1996.

Dibdin, Thomas Frognall. *Reminiscences of a Literary Life*. London: J. Major, 1836.

Dolan, Brian. "Conservative Politicians, Radical Philosophers and the Aerial Remedy for the Diseases of Civilization." *History of the Human Sciences* 15, no. 2 (2002): 35–54.

Donovan, Arthur L. *Philosophical Chemistry in the Scottish Enlightenment: The Doctrines and Discoveries of William Cullen and Joseph Black*. Edinburgh: Edinburgh University Press, 1975.

Douglas, Aileen. "Popular Science and the Representation of Women: Fontenelle and After." *Eighteenth-Century Life* 18, no. 2 (1994): 1–14.

Eliot, George. *Middlemarch*. Edited by W. J. Harvey. Harmondsworth, UK: Penguin Books, 1985. Originally published 1871–72.

Elliott, Paul. "'More Subtle than the Electric Aura': Georgian Medical Electricity, the Spirit of Animation and the Development of Erasmus Darwin's Psychophysiology." *Medical History* 52 (2008): 195–220.

Elsner, Jas, and Joan-Pau Rubiés, eds. *Voyages and Visions: Towards a Cultural History of Travel*. London: Reaktion Books, 1999.

Fara, Patricia. *Newton: The Making of Genius*. London: Macmillan, 2002.

Faraday, Michael. *Curiosity Perfectly Satisfyed: Faraday's Travels in Europe 1813–1815.* Edited by Brian Bowers and Lenore Symons. London: Peter Peregrinus Ltd. in association with the Science Museum, 1991.

Ferriar, John. Review of *Notice of Some Observations Made at the Medical Pneumatic Institution,* by Thomas Beddoes. *Monthly Review,* 2nd ser., 30 (September–December 1799): 60–72.

Finnegan, Diarmid A. "The Spatial Turn: Geographical Approaches in the History of Science." *Journal of the History of Biology* 41 (2008): 369–88.

Forgan, Sophie, ed. *Science and the Sons of Genius: Studies on Humphry Davy.* London: Science Reviews, 1980.

Foucault, Michel. *The History of Sexuality.* Vol. 1, *An Introduction.* Vol. 2, *The Use of Pleasure.* Vol. 3, *The Care of the Self.* New York: Vintage Books, 1990.

Fulford, Tim. *Romanticism and Masculinity: Gender, Politics, and Poetics in the Writings of Burke, Coleridge, Cobbett, Wordsworth, De Quincey, and Hazlitt.* Basingstoke, UK: Macmillan, 1999.

———. "Conducting the Vital Fluid: The Politics and Poetics of Mesmerism in the 1790s." *Studies in Romanticism* 43, no. 1 (2004): 57–78.

Fullmer, June Z. "Letter." *Scientific American* 203, no. 2 (August 1960): 12–14.

———. "Humphry Davy's Adversaries." *Chymia* 8 (1962): 147–64.

———. "Humphry Davy and the Gunpowder Manufactory." *Annals of Science* 20 (1964): 165–94.

———. "Davy's Biographers: Notes on Scientific Biography." *Science* 155, no. 3760 (20 January 1967): 285–91.

———. "Davy's Sketches of His Contemporaries." *Chymia* 12 (1967): 127–50.

———. *Sir Humphry Davy's Published Works.* Cambridge, MA: Harvard University Press, 1969.

———. "Humphry Davy, Reformer." In Forgan, *Science and the Sons of Genius,* 59–94.

———. *Young Humphry Davy: The Making of an Experimental Chemist.* Philadelphia: American Philosophical Society, 2000.

Funnell, Peter. "Lawrence among Men: Friends, Patrons and the Male Portrait." In *Thomas Lawrence: Regency Power and Brilliance,* edited by Cassandra Albinson, Peter Funnell, and Lucy Peltz, 1–25. New Haven, CT: Yale Center for British Art; London: National Portrait Gallery, 2011.

Fyfe, Aileen, and Bernard Lightman. "Science in the Marketplace: An Intro-duction." In Fyfe and Lightman, *Science in the Marketplace*, 1–19.

———, eds. *Science in the Marketplace: Nineteenth-Century Sites and Experiences*. Chicago: University of Chicago Press, 2007.

"G. M., Dr." "On the State of Galvanism and Other Scientific Pursuits in Germany." *[Nicholson's] Journal of Natural Philosophy, Chemistry and the Arts*, 1st ser. , 4 (1800–01): 511–13.

Garfinkle, Norton. "Science and Religion in England, 1790–1800: The Critical Response to the Work of Erasmus Darwin." *Journal of the History of Ideas* 16 (1955): 376–88.

Gascoigne, John. *Joseph Banks and the English Enlightenment*. Cambridge: Cambridge University Press, 1994.

Gaukroger, Stephen. "Biography as a Route to Understanding Early Modern Natural Philosophy." In Söderqvist, *History and Poetics of Scientific Biography*, 37–49.

Gibbes, George Smith. *A Phlogistic Theory Ingrafted upon M. Fourcroy's Philosophy of Chemistry*. Bath: W. Meyler and Son, 1809.

Gilbert, Arthur N. "Sexual Deviance and Disaster during the Napoleonic Wars." *Albion* 9 (1977): 98–113.

Gilroy, Amanda, ed. *Romantic Geographies: Discourses of Travel, 1775–1844*. Manchester, UK: Manchester University Press, 2000.

Golinski, Jan. *Science as Public Culture: Chemistry and Enlightenment in Britain, 1760–1820*. Cambridge: Cambridge University Press, 1992.

———. "'The Nicety of Experiment': Precision of Measurement and Precision of Reasoning in Late Eighteenth-Century Chemistry." In *The Values of Precision*, edited by M. Norton Wise, 72–91. Princeton, NJ: Princeton University Press, 1995.

———. "The Care of the Self and the Masculine Birth of Science." *History of Science* 40 (2002): 125–45.

Greenblatt, Stephen. *Renaissance Self-Fashioning: From More to Shakespeare*. Chicago: University of Chicago Press, 1980.

Griggs, Earl Leslie, ed. *Collected Letters of Samuel Taylor Coleridge*. 6 vols. Oxford: Clarendon Press, 1956–71.

Hall, Marie Boas. *All Scientists Now: The Royal Society in the Nineteenth Century*. Cambridge: Cambridge University Press, 1984.

Hannaway, Owen. *The Chemists and the Word: The Didactic Origins of Chemistry*. Baltimore: Johns Hopkins University Press, 1975.

Haraway, Donna J. "Modest Witness: Feminist Diffractions in Science Studies." In *The Disunity of Science: Boundaries, Contexts, and Power*, edited by Peter Galison and David J. Stump, 428–41. Stanford, CA: Stanford University Press, 1996.

Hartley, Harold. *Humphry Davy*. Wakefield, UK: EP Publishing, 1972.

Haygarth, John. *Of the Imagination as a Cause and as a Cure of Disorders of the Body: Exemplified by Fictitious Tractors, and Epidemical Convulsions*. Bath, UK: R. Cruttwell, 1800.

Hays, J. N. "The London Lecturing Empire, 1800–50." In *Metropolis and Province: Science in British Culture, 1780–1850*, edited by Ian Inkster and Jack Morrell, 91–119. Philadelphia: University of Pennsylvania Press, 1983.

Hazlitt, William. "The Indian Jugglers." In *The Fight and Other Writings*, edited by Tom Paulin and David Chandler, 114–28. London: Penguin Books, 2000. Originally published 1821.

Henry, William. *Elements of Experimental Chemistry*. 11th ed. 2 vols. London: Baldwin and Cradock, 1829.

Herzig, Rebecca M. *Suffering for Science: Reason and Sacrifice in Modern America*. New Brunswick, NJ: Rutgers University Press, 2005.

Heyd, Michael. "The Reaction to Enthusiasm in the Seventeenth Century: Towards an Integrative Approach." *Journal of Modern History* 53, no. 2 (1981): 258–80.

Higgitt, Rebekah. "Discriminating Days? Partiality and Impartiality in Nineteenth-Century Biographies of Newton." In Söderqvist, *History and Poetics of Scientific Biography*, 155–72.

Hitchcock, Tim, and Michèle Cohen, eds. *English Masculinities, 1600–1800*. London: Longman, 1999.

Hodges, Andrew. *Alan Turing: The Enigma of Intelligence*. London: Burnett Books, 1983.

Holland, Lady Elizabeth Vassall Fox. *The Journal of Elizabeth Lady Holland (1791–1811)*. Edited by Giles Stephen Holland Fox-Strangways, Earl of Ilchester. 2 vols. London: Longmans, Green, 1908.

Holmes, Frederic L. "The 'Revolution in Chemistry and Physics': Overthrow of a Reigning Paradigm or Competition between Contemporary Research Programs?" *Isis* 91 (2000): 735–53.

Holmes, Richard. *The Age of Wonder: How the Romantic Generation Discovered the Beauty and Terror of Science*. London: HarperPress, 2008.

———. *Thomas Lawrence Portraits*. London: National Portrait Gallery, 2010.

Hoover, Suzanne R. "Coleridge, Humphry Davy, and Some Early Experiments with a Consciousness-Altering Drug." *Bulletin of Research in the Humanities* 81 (1978): 9–27.

Houghton, Walter E., ed. *The Wellesley Index to Victorian Periodicals, 1824–1900*. 5 vols. Toronto: University of Toronto Press, 1966–87.

Humboldt, Alexander von. *Views of Nature; or, Contemplations on the Sublime Phenomena of Creation*. Translated by E. C. Otté and Henry G. Bohn. London: Bohn, 1850.

Hunt, Lynn, and Margaret C. Jacob. "The Affective Revolution in 1790s Britain." *Eighteenth-Century Studies* 34, no. 4 (2001): 491–521.

Inkster, Ian. "Science and Society in the Metropolis: A Preliminary Examination of the Social and Institutional Context of the Askesian Society of London, 1796–1807." *Annals of Science* 34 (1977): 1–32.

———. "The Public Lecture as an Instrument of Science Education for Adults—The Case of Great Britain, c. 1750–1850." *Paedagogica Historica* 20 (1980): 80–107.

Jacob, Margaret C., and Michael J. Sauter. "Why Did Humphry Davy and Associates Not Pursue the Pain-Alleviating Effects of Nitrous Oxide?" *Journal of the History of Medicine and Allied Sciences* 57, no. 2 (2002): 161–76.

Jacyna, L. S. "Immanence or Transcendence: Theories of Life and Organization in Britain, 1790–1835," *Isis* 74 (1983): 310–29.

James, Frank A. J. L. "Davy in the Dockyard: Humphry Davy, the Royal Society and the Electro-Chemical Protection of the Copper Sheeting of His Majesty's Ships in the Mid 1820s." *Physis* 29, no. 1 (1992): 205–25.

———, ed. *"The Common Purposes of Life": Science and Society at the Royal Institution of Great Britain*. Aldershot, UK: Ashgate, 2002.

———. "How Big Is a Hole?: The Problems of the Practical Application of Science in the Invention of the Miners' Safety Lamp by Humphry Davy and George Stephenson in Late Regency England." *Transactions of the Newcomen Society* 75 (2005): 175–227.

———. *Michael Faraday: A Very Short Introduction*. Oxford: Oxford University Press, 2010.

James, Frank A. J. L., and Anthony Peers. "Constructing Space for Science at the Royal Institution of Great Britain." *Physics in Perspective* 9, no. 2 (2007): 130–85.

Janković, Vladimir. *Reading the Skies: A Cultural History of English Weather, 1650–1820*. Chicago: University of Chicago Press, 2000.

Jay, Mike. *The Atmosphere of Heaven: The Unnatural Experiments of Dr. Beddoes and His Sons of Genius*. New Haven, CT: Yale University Press, 2009.

———. "The Atmosphere of Heaven: The 1799 Nitrous Oxide Researches Reconsidered." *Notes and Records of the Royal Society* 63, no. 3 (2009): 297–309.

Johns, Adrian. *The Nature of the Book: Print and Knowledge in the Making*. Chicago: University of Chicago Press, 1998.

Johns-Putra, Adeline. "'Blending Science with Literature': The Royal Institution, Eleanor Anne Porden and The Veils." *Nineteenth-Century Contexts* 33, no. 1 (2011): 35–52.

Jones, Henry Bence. *The Royal Institution: Its Founder and Its First Professors*. London: Longmans, Green, 1871.

Jones, Matthew L. *The Good Life in the Scientific Revolution: Descartes, Pascal, Leibniz, and the Cultivation of Virtue*. Chicago: University of Chicago Press, 2006.

Jonsson, Fredrik Albritton. *Enlightenment's Frontier: The Scottish Highlands and the Origins of Environmentalism*. New Haven, CT: Yale University Press, 2013.

Jordanova, Ludmilla. *Sexual Visions: Images of Gender in Science and Medicine between the Eighteenth and Twentieth Centuries*. New York: Harvester Wheatsheaf, 1989.

———. "Sex and Gender." In *Inventing Human Science: Eighteenth-Century Domains*, edited by Christopher Fox, Roy Porter, and Robert Wokler, 152–83. Berkeley: University of California Press, 1995.

Keller, Evelyn Fox. *Reflections on Gender and Science*. New Haven, CT: Yale University Press, 1985.

Kenyon, T. K. "Science and Celebrity: Humphry Davy's Rising Star." *Chemical Heritage* 26, no. 4 (2009): 30–35.

Kim, Mi Gyung. *Affinity, That Elusive Dream: A Genealogy of the Chemical Revolution*. Cambridge, MA: MIT Press, 2003.

———. "The 'Instrumental' Reality of Phlogiston." *HYLE–International Journal for Philosophy of Chemistry* 14, no. 1 (2008): 27–51.

Kipnis, Naum. "Luigi Galvani and the Debate on Animal Electricity." *Annals of Science* 44 (1987): 107–42.

Klancher, Jon P. *Transfiguring the Arts and Sciences: Knowledge and Cultural Institutions in the Romantic Age*. Cambridge: Cambridge University Press, 2013.

Klein, Lawrence E. "Sociability, Solitude, and Enthusiasm." *Huntington Library Quarterly* 60, no. 1/2 (1997): 153–77.

Knight, David. "Accomplishment or Dogma: Chemistry in the Introductory Works of Jane Marcet and Samuel Parkes." *Ambix* 33 (1986): 94–98.

———. *Humphry Davy: Science and Power*. Oxford: Basil Blackwell, 1992.

———. "Establishing the Royal Institution: Rumford, Banks and Davy." In James, *"Common Purposes of Life,"* 97–118.

———. "Chemists Get Down to Earth." In *The Making of the Geological Society of London*, edited by C. L. E. Lewis and S. J. Knell, 93–103. London: Geological Society of London, 2009.

Knott, Sarah, and Barbara Taylor, eds. *Women, Gender, and Enlightenment*. Basingstoke, UK: Palgrave Macmillan, 2005.

Kuhn, Thomas S. " Mathematical versus Experimental Traditions in the Development of Physical Science." In *The Essential Tension: Selected Studies in Scientific Tradition and Change*, edited by Thomas Kuhn, 31–65. Chicago: University of Chicago Press, 1977.

Kurzer, Frederick. "William Hasledine Pepys FRS: A Life in Scientific Research, Learned Societies and Technical Enterprise." *Annals of Science* 60, no. 2 (2003): 137–83.

Lamont-Brown, Raymond. *Humphry Davy: Life beyond the Lamp*. Stroud, UK: Sutton, 2004.

Laqueur, Thomas. *Making Sex: Body and Gender from the Greeks to Freud*. Cambridge, MA: Harvard University Press, 1990.

Leask, Nigel. *Curiosity and the Aesthetics of Travel Writing, 1770–1840: "From an Antique Land."* Oxford: Oxford University Press, 2002.

Lefebure, Molly. "Humphry Davy: Philosophic Alchemist." In *The Coleridge Connection: Essays for Thomas McFarland*, edited by Richard Gravil and Molly Lefebure, 82–106. London: Macmillan, 1990.

Levere, Trevor H. *Poetry Realized in Nature: Samuel Taylor Coleridge and Early Nineteenth-Century Science*. Cambridge: Cambridge University Press, 1981.

———. "Dr. Thomas Beddoes at Oxford: Radical Politics in 1788–1793 and the Fate of the Regius Chair in Chemistry." *Ambix* 28 (1981): 61–69.

———. "Dr. Thomas Beddoes (1760–1808) and the Lunar Society of Birmingham: Collaborations in Medicine and Science." *British Journal for Eighteenth-Century Studies* 30 (2007): 209–26.

Lightman, Bernard. "Refashioning the Spaces of London Science: Elite Epistemes in the Nineteenth Century." In *Geographies of Nineteenth-Century Science*, edited by David N. Livingstone and Charles W. J. Withers, 25–50. Chicago: University of Chicago Press, 2011.

Lindee, M. Susan. "The American Career of Jane Marcet's *Conversations on Chemistry*, 1806–1853." *Isis* 82 (1991): 8–23.

Lucier, Paul. "The Professional and the Scientist in Nineteenth-Century America," *Isis* 100 (2009): 699–732.

Lucretius. *On the Nature of Things*. Edited and translated by Anthony M. Esolen. Baltimore: Johns Hopkins University Press, 1995.

Lyell, Charles. *Principles of Geology: Being An Attempt to Explain the Former Changes of the Earth's Surface by Reference to Causes Now in Operation*. 3 vols. London: John Murray, 1830–33.

———. *Principles of Geology*. Edited with an introduction by James A. Secord. London: Penguin Books, 1997.

Mac Arthur, C. W. P. "Davy's Differences with Gay-Lussac and Thenard: New Light on Events in Paris and on the Transmission and Translation of Davy's Papers in 1810." *Notes and Records of the Royal Society of London* 39, no. 2 (1985): 207–28.

Macaulay, Catharine. *Letters on Education: With Observations on Religious and Metaphysical Subjects*. London: C. Dilly, 1790.

MacLeod, Christine. "James Watt, Heroic Invention and the Idea of the Industrial Revolution." In *Technological Revolution in Europe*, edited by Maxine Berg and Kristine Bruland, 96–116. Cheltenham, UK: Edward Elgar, 1998.

———. *Heroes of Invention: Technology, Liberalism and British Identity, 1750–1914*. Cambridge: Cambridge University Press, 2007.

———. "Distinguished Men of Science of Great Britain Living in 1807–8." *Oxford Dictionary of National Biography* online edition. Accessed 13 January 2015. http://www.oxforddnb.com/view/theme/97115.

Macnish, Robert. *The Anatomy of Drunkenness*. 4th ed. Glasgow: W. R. M'Phun, 1832.

[Marcet, Jane.] *Conversations on Chemistry in which the Elements of that Science are Familiarly Explained and Illustrated by Experiments*. 2 vols. London: Longman, Hurst, Rees, Orme, and Brown, 1806.

Martin, Raymond, and John Baressi. *Naturalization of the Soul: Self and Personal Identity in the Eighteenth Century*. London: Routledge, 2000.

Martin, Thomas. "Origins of the Royal Institution." *British Journal for the History of Science* 1, no. 1 (1962): 49–63.

———. "Presidential Address: Early Years at the Royal Institution." *British Journal for the History of Science* 2, no. 2 (1964): 99–115.

Martineau, Harriet. *The History of England During the Thirty Years' Peace, 1816–1846*. 2 vols. London: Charles Knight, 1849–50.

Mazzotti, Massimo. "Newton for Ladies: Gentility, Gender and Radical Culture." *British Journal for the History of Science* 37, no. 2 (2004): 119–46.

McCalman, Iain. *Radical Underworld: Prophets, Revolutionaries, and Pornographers in London, 1795–1840*. Cambridge: Cambridge University Press, 1988.

McMahon, Darrin M. *Divine Fury: A History of Genius*. New York: Basic Books, 2013.

Mee, John. "Anxieties of Enthusiasm: Coleridge, Prophecy, and Popular Politics in the 1790s." *Huntington Library Quarterly* 60, no. 1/2 (1997): 179–203.

———. *Romanticism, Enthusiasm, and Regulation: Poetics and the Policing of Culture in the Romantic Period*. Oxford: Oxford University Press, 2003.

Melhado, Evan M. "Chemistry, Physics, and the Chemical Revolution." *Isis* 76, no. 2 (1985): 195–211.

Mellor, Anne K. *Mary Shelley: Her Life, Her Fiction, Her Monsters*. New York: Routledge, 1989.

Merton, Robert K. *The Sociology of Science: Theoretical and Empirical Investigations*. Edited by Norman W. Storer. Chicago: University of Chicago Press, 1973.

Metzner, Paul. *Crescendo of the Virtuoso: Spectacle, Skill, and Self-Promotion in Paris during the Age of Revolution*. Berkeley: University of California Press, 1998.

Mialet, Hélène. *Hawking Incorporated: Stephen Hawking and the Anthropology of the Knowing Subject*. Chicago: University of Chicago Press, 2012.

Miles, Wyndham D. "Sir Humphrey Davie, The Prince of Agricultural Chemists." *Chymia* 7 (1961): 126–34.

Miller, David Philip. "Between Hostile Camps: Sir Humphry Davy's Presidency of the Royal Society of London, 1820–1827." *British Journal for the History of Science* 16, no. 1 (1983): 1–47.

———. "'Into the Valley of Darkness': Reflections on the Royal Society in the Eighteenth Century." *History of Science* 27 (1989): 155–66.

———. "The Usefulness of Natural Philosophy: The Royal Society and the Culture of Practical Utility in the Later Eighteenth Century." *British Journal for the History of Science* 32, no. 2 (1999): 185–201.

———. "'Puffing Jamie': The Commercial and Ideological Importance of Being a 'Philosopher' in the Case of the Reputation of James Watt (1736–1819)." *History of Science* 38 (2000): 1–24.

———. *James Watt, Chemist: Understanding the Origins of the Steam Age*. London: Pickering and Chatto, 2009.

———. "Mannered Science and Political Identity." *Metascience* 19, no. 1 (2010): 133–35.

Miller, David Philip, and Trevor H. Levere. "'Inhale It and See?' The Collaboration between Thomas Beddoes and James Watt in Pneumatic Medicine." *Ambix* 55, no. 1 (2008): 5–28.

Mitchell, Robert. *Experimental Life: Vitalism in Romantic Science and Literature*. Baltimore: Johns Hopkins University Press, 2013.

Moers, Ellen. *The Dandy: Brummell to Beerbohm*. New York: Viking, 1960.

Moran, Bruce T. *Distilling Knowledge: Alchemy, Chemistry, and the Scientific Revolution*. Cambridge, MA: Harvard University Press, 2005.

Morrell, Jack B. "Professionalisation." In *Companion to the History of Mod-*

ern Science, edited by Robert C. Olby, et al., 980–89. London: Routledge, 1990.

Morrell, Jack, and Arnold Thackray. *Gentlemen of Science: Early Years of the British Association for the Advancement of Science*. Oxford: Clarendon Press, 1981.

Morus, Iwan Rhys. *Frankenstein's Children: Electricity, Exhibition, and Experiment in Early-Nineteenth-Century London*. Princeton, NJ: Princeton University Press, 1998.

———. "'More the Aspect of Magic than Anything Natural': The Philosophy of Demonstration." In Fyfe and Lightman, *Science in the Marketplace*, 336–70.

———. "Radicals, Romantics and Electrical Showmen: Placing Galvanism at the End of the English Enlightenment." *Notes and Records of the Royal Society* 63, no. 3 (2009): 263–75.

Morus, Iwan, Simon Schaffer, and Jim Secord. "Scientific London." In *London—World City, 1800–1840*, edited by Celina Fox, 129–42. New Haven, CT: Yale University Press, 1992.

Murray, John. *System of Chemistry*. 4th ed. 4 vols. Edinburgh: Francis Pillans, 1819.

Myers, Greg. "Jane Marcet's *Conversations on Chemistry*: Fictionality, Demonstration, and a Forum for Popular Science." In *Natural Eloquence: Women Reinscribe Science*, edited by Barbara Gates and Ann Shteir, 43–60. Madison: University of Wisconsin Press, 1997.

Nicholson, Malcolm. "Alexander von Humboldt and the Geography of Vegetation." In Cunningham and Jardine, *Romanticism and the Sciences*, 169–85.

Nickles, Thomas. "Discovery." In *Companion to the History of Modern Science*, edited by Robert C. Olby et al., 148–65. London: Routledge, 1990.

Ørsted, Hans Christian. *The Soul in Nature: With Supplementary Contributions*. Translated by Leonora Horner and Joanna B. Horner. London: Dawsons of Pall Mall, 1966. Originally published 1852.

Otto, Peter. "Performing the Resurrection: James Graham and the Multiplication of the Real." *Cultural & Social History* 3, no. 3 (2006): 325–40.

Paine, Thomas. *The Rights of Man*. Harmondsworth, UK: Penguin Books, 1984. Originally published 1791.

Pancaldi, Giuliano. *Volta: Science and Culture in the Age of Enlightenment.* Princeton, NJ: Princeton University Press, 2003.

———. "On Hybrid Objects and Their Trajectories: Beddoes, Davy and the Battery." *Notes and Records of the Royal Society* 63, no. 3 (2009): 247–62.

Parker, W. M. "Lady Davy in Her Letters." *Quarterly Review* 300 (1962): 79–89.

Paris, John Ayrton. *The Life of Sir Humphry Davy.* London: Henry Colburn and Richard Bentley, 1831. (Also published in two volumes.)

Paton-Williams, David. *Katterfelto: Prince of Puff.* Leicester: Matador, 2008.

Pera, Marcello. *The Ambiguous Frog: The Galvani-Volta Controversy on Animal Electricity.* Translated by Jonathan Mandelbaum. Princeton, NJ: Princeton University Press, 1992.

Pocock, J. G. A. "Enthusiasm: The Antiself of Enlightenment." *Huntington Library Quarterly* 60, no. 1/2 (1997): 7–28.

Polkinghorn, Bette. *Jane Marcet: An Uncommon Woman.* Aldermaston, UK: Forestwood, 1993.

Polwhele, Richard. *The Unsex'd Females: A Poem Addressed to the Author of The Pursuits of Literature.* London: Cadell and Davies, 1798.

Porter, Roy S. "Gentlemen and Geology: The Emergence of a Scientific Career, 1660–1920." *Historical Journal* 21 (1978): 809–36.

———. "The Sexual Politics of James Graham." *British Journal for Eighteenth-Century Studies* 5 (1982): 199–206.

———. *Doctor of Society: Thomas Beddoes and the Sick Trade in Late-Enlightenment Britain.* London: Routledge, 1992.

Porter, Theodore M. *Karl Pearson: The Scientific Life in a Statistical Age.* Princeton, NJ: Princeton University Press, 2006.

———. "Is the Life of the Scientist a Scientific Unit?" *Isis* 97 (2006): 314–21.

Powell, Richard C. "Geographies of Science: Histories, Localities, Practices, Futures." *Progress in Human Geography* 31, no. 3 (2007): 309–29.

Pratt, Mary Louise. *Imperial Eyes: Travel Writing and Transculturation.* London: Routledge, 1992.

Prescott, G. M. "Forging Identity: The Royal Institution's Visual Collections." In James, *"Common Purposes of Life,"* 59–96.

Priestley, Joseph. "Observations and Experiments Relating to the Pile of Volta." *Nicholson's Journal*, 2nd ser., 1 (March 1802): 198–204.

Rendall, Jane. "'Women That Would Plague Me with Rational Conversation': Aspiring Women and Scottish Whigs, c. 1790–1830." In Knott and Taylor, *Women, Gender and Enlightenment*, 326–47.

Reiss, Timothy J. *Mirages of the Selfe: Patterns of Personhood in Ancient and Early Modern Europe*. Stanford, CA: Stanford University Press, 2003.

Richardson, Alan. *British Romanticism and the Science of the Mind*. Cambridge: Cambridge University Press, 2001.

Roberts, Lissa, Simon Schaffer, and Peter Dear, eds. *The Mindful Hand: Inquiry and Invention from the Late Renaissance to Early Industrialisation*. Amsterdam: KNAW, 2007.

Robinson, Eric, and Douglas McKie. *Partners in Science: Letters of James Watt and Joseph Black*. Cambridge, MA: Harvard University Press, 1970.

Ross, Sydney. "Scientist: The Story of a Word." *Annals of Science* 18, no. 2 (1962): 65–85.

Rossi, Paolo. *The Dark Abyss of Time: The History of the Earth and the History of Nations from Hooke to Vico*. Translated by Lydia G. Cochrane. Chicago: University of Chicago Press, 1984.

Rossotti, Hazel. Introduction to *Chemistry in the Schoolroom: 1806; Selections from Mrs. Marcet's "Conversations on Chemistry,"* edited by Hazel Rossotti, i–xxi. Bloomington, IN: Author House, 2006.

Rudwick, Martin J. S. "The Emergence of a Visual Language for Geological Science." *History of Science* 14 (1976): 149–95.

———. *Bursting the Limits of Time: The Reconstruction of Geohistory in the Age of Revolution*. Chicago: University of Chicago Press, 2005.

———. *Worlds Before Adam: The Reconstruction of Geohistory in the Age of Reform*. Chicago: University of Chicago Press, 2008.

Russell, Colin A. *Science and Social Change, 1700–1900*. London: Macmillan, 1983.

Ruston, Sharon. *Creating Romanticism: Case Studies in the Literature, Science and Medicine of the 1790s*. Basingstoke, UK: Palgrave Macmillan, 2013.

Sandford, Margaret Poole. *Thomas Poole and His Friends*, 2 vols. London and New York: Macmillan, 1888.

Sarafianos, Aris. "The Contractility of Burke's Sublime and Heterodoxies in Medicine and Art." *Journal of the History of Ideas* 69:1 (2007): 23–48.

Schaffer, Simon. "Scientific Discoveries and the End of Natural Philosophy." *Social Studies of Science* 16, no. 3 (1986): 387–420.

———. "Genius in Romantic Natural Philosophy." In Cunningham and Jardine, *Romanticism and the Sciences*, 82–98.

———. "Self Evidence." *Critical Inquiry* 18, no. 2 (1992): 327–62.

———. "Making up Discovery." In *Dimensions of Creativity*, edited by Margaret A. Boden, 13–51. Cambridge, MA: MIT Press, 1994.

———. "The Astrological Roots of Mesmerism." *Studies in History and Philosophy of Science (Part C)* 41, no. 2 (2010): 158–68.

Schelling, F. W. J. *Clara; or, On Nature's Connection to the Spirit World*. Translated by Fiona Steinkamp. Albany, NY: State University of New York Press, 2002. Originally published 1861.

Schiebinger, Londa. *The Mind Has No Sex? Women in the Origins of Modern Science*. Cambridge, MA: Harvard University Press, 1989.

Scott, Walter. *The Journal of Sir Walter Scott, 1825–32, from the Original Manuscript at Abbotsford*. New ed. Edinburgh: David Douglas, 1891.

Secord, James A. "Knowledge in Transit." *Isis* 95 (2004): 654–72.

———. *Visions of Science: Books and Readers at the Dawn of the Victorian Age*. Oxford: Oxford University Press, 2014.

Sennett, Richard. *The Fall of Public Man*. Cambridge: Cambridge University Press, 1974.

Shapin, Steven. "Pump and Circumstances: Robert Boyle's Literary Technology." *Social Studies of Science* 14 (1984): 481–520.

———. *A Social History of Truth: Civility and Science in Seventeenth-Century England*. Chicago: University of Chicago Press, 1994.

———. "The Image of the Man of Science." In *The Cambridge History of Science*, vol. 4, *Eighteenth-Century Science*, edited by Roy Porter, 159–83. Cambridge: Cambridge University Press, 2003.

———. *The Scientific Life: A Moral History of a Late Modern Vocation*. Chicago: University of Chicago Press, 2008.

Shapin, Steven, and Simon Schaffer. *Leviathan and the Air-Pump: Hobbes, Boyle, and the Experimental Life*. 2nd ed. Princeton, NJ: Princeton University Press, 2011.

Shelley, Mary. *Frankenstein; or, The Modern Prometheus*. 1818 Text. Edited by Marilyn Butler. Oxford: Oxford University Press, 1994.

Sher, Richard B. *The Enlightenment and the Book: Scottish Authors and their Publishers in Eighteenth-Century Britain, Ireland, and America.* Chicago: University of Chicago Press, 2006.

Shortland, Michael, and Richard Yeo, eds. *Telling Lives in Science: Essays on Scientific Biography.* Cambridge: Cambridge University Press, 1996.

Siegfried, Robert. "Boscovich and Davy: Some Cautionary Remarks." *Isis* 58 (1967): 236–38.

Siegfried, Robert, and Robert H. Dott. "Humphry Davy as Geologist, 1805–29." *British Journal for the History of Science* 9, no. 2 (1976): 219–27.

Silliman, Benjamin. *A Journal of Travels in England, Holland, and Scotland: And of Two Passages Over the Atlantic, in the Years 1805 and 1806,* 2nd ed. 2 vols. Boston: T. B. Wait, 1812.

Simond, Louis. *Journal of a Tour and Residence in Great Britain, During the Years 1810 and 1811,* 2nd ed. 2 vols. Edinburgh: J. Ballantyne, 1817.

Smiles, Samuel. *Self-Help: With Illustrations of Character and Conduct.* London: John Murray, 1859.

———, ed. *A Publisher and His Friends: Memoir and Correspondence of the Late John Murray, With An Account of the Origin and Progress of the House, 1768–1843.* 2 vols. London: John Murray, 1891.

Smith, Roger. "Self-Reflection and the Self." In *Rewriting the Self: Histories from the Middle Ages to the Present,* edited by Roy Porter, 49–57. London: Routledge, 1997.

Snyder, Laura J. *The Philosophical Breakfast Club: Four Remarkable Friends Who Transformed Science and Changed the World.* New York: Broadway Books, 2011.

Söderqvist, Thomas. "'No Genre of History Fell under More Odium Than That of Biography': The Delicate Relations between Scientific Biography and the Historiography of Science." In Söderqvist, *History and Poetics of Scientific Biography,* 242–62.

———, ed. *History and Poetics of Scientific Biography.* Abingdon, UK: Ashgate, 2007.

Soper, Kate. "Feminism and Enlightenment Legacies." In Knott and Taylor, *Women, Gender and Enlightenment,* 705–15.

[Southey, Robert.] *Letters from England by Don Manuel Alvarez Espriella.* New York: G. Dearborn, 1836.

Stansfield, Dorothy A., and Ronald G. Stansfield. *Thomas Beddoes, M.D., 1760–1808: Chemist, Physician, Democrat.* Dordrecht, Netherlands: D. Reidel, 1984.

Steigerwald, Joan. "The Subject as Instrument: Galvanic Experiments, Organic Apparatus and Problems of Calibration." In *Human Experimentation*, edited by Larry Stewart and E. Dyck. Forthcoming.

Stewart, Larry. "His Majesty's Subjects: From Laboratory to Human Experiment in Pneumatic Chemistry." *Notes and Records of the Royal Society* 63, no. 3 (2009): 231–45.

Stewart, Larry, and Paul Weindling. "Philosophical Threads: Natural Philosophy and Public Experiment among the Weavers of Spitalfields." *British Journal for the History of Science* 28, no. 1 (1995): 37–62.

Stone, Lawrence. *The Family, Sex and Marriage in England 1500–1800.* Abridged ed. Harmondsworth, UK: Penguin Books, 1979.

Strickland, Stuart W. "Galvanic Disciplines: The Boundaries, Objects, and Identities of Experimental Science in the Era of Romanticism." *History of Science* 33 (1995): 449–68.

———. "The Ideology of Self-Knowledge and the Practice of Self-Experimentation." *Eighteenth-Century Studies* 31, no. 4 (1998): 453–71.

Sudduth, William M. "Eighteenth-Century Identifications of Electricity with Phlogiston." *Ambix* 25 (1978): 131–47.

———. "The Voltaic Pile and Electro-chemical Theory in 1800." *Ambix* 27 (1980): 26–35.

Sumner, James. "Michael Combrune, Peter Shaw and Commercial Chemistry: The Boerhaavian Chemical Origins of Brewing Thermometry." *Ambix* 54 (2007): 5–29.

Sutton, Geoffrey. "The Politics of Science in Early Napoleonic France: The Case of the Voltaic Pile." *Historical Studies in the Physical Sciences* 11 (1981): 329–66.

———. *Science for a Polite Society: Gender, Culture, and the Demonstration of Enlightenment.* Boulder, CO: Westview Press, 1995.

Sylvester, Charles. *An Elementary Treatise on Chemistry.* Liverpool: E. and W. Smith, 1809.

Taylor, Charles. *Sources of the Self: The Making of the Modern Identity.* Cambridge, MA: Harvard University Press, 1989.

Terrall, Mary. "Biography as Cultural History of Science." *Isis* 97 (2006): 306–13.

Thompson, Carl. *The Suffering Traveller and the Romantic Imagination.* Oxford: Clarendon Press, 2007.

[Thomson, Thomas.] Review of *Elements of Chemical Philosophy.* by Humphry Davy. *Annals of Philosophy* 1 (May 1813): 371–77.

———. *The History of Chemistry.* 2 vols. London: Henry Colburn and Richard Bentley, 1830–31.

Thorpe, T. E. *Humphry Davy: Poet and Philosopher.* 1896. Reprint, Stroud, UK: Nonsuch, 2007.

Tobin, J. J. *Journal of a Tour Made in the Years 1828 and 1829 through Styria, Carniola, and Italy, Whilst Accompanying the Late Sir Humphry Davy.* London: W. S. Orr, 1832.

Treneer, Anne. *The Mercurial Chemist: A Life of Sir Humphry Davy.* London: Methuen, 1963.

Tresch, John. *The Romantic Machine: Utopian Science and Technology after Napoleon.* Chicago: University of Chicago Press, 2012.

Trittel, Rebecca Blass. "Genius on Canvas: The Portraiture of Thomas Phillips, R.A. (1770–1845)." PhD diss., University of Essex, 2005.

Tucker, Abraham. *The Light of Nature Pursued, By Edward Search, Esq.* 5 vols. London: T. Jones, 1768.

Uglow, Jenny. *The Lunar Men: Five Friends Whose Curiosity Changed the World.* New York: Farrar, Straus and Giroux, 2002.

Unwin, Patrick, and Robert Unwin. "'A Devotion to the Experimental Sciences and Arts': The Subscription to the Great Battery at the Royal Institution 1808-9." *British Journal for the History of Science* 40, no. 2 (2007): 181–203.

———. "Humphry Davy and the Royal Institution of Great Britain." *Notes and Records of the Royal Society* 63, no. 1 (2009): 7–33.

Ure, Andrew. "Experiments on the Relation between Muriatic Acid and Chlorine." *Transactions of the Royal Society of Edinburgh* 8 (1818): 329–53.

Vermeir, Koen, and Michael Funk Deckard, eds. *The Science of Sensibility: Reading Burke's Philosophical Enquiry.* Dordrecht, Netherlands: Springer, 2011.

Vickery, Amanda. *The Gentleman's Daughter: Women's Lives in Georgian England*. New Haven, CT: Yale University Press, 1998.

Wahrman, Dror. *The Making of the Modern Self: Identity and Culture in Eighteenth-Century England*. New Haven, CT: Yale University Press, 2004.

Walls, Laura Dassow. *The Passage to Cosmos: Alexander von Humboldt and the Shaping of America*. Chicago: University of Chicago Press, 2009.

Walker, Ezekiel. *Philosophical Essays Selected from the Originals Printed in the Philosophical Journals*. Lynn, UK: John Wade, 1823.

Walker, Richard. *Regency Portraits*. 2 vols. London: National Portrait Gallery, 1985.

Walker, William Jr. *Memoirs of the Distinguished Men of Science of Great Britain Living in the Years 1807-8*. London: W. Walker, 1862.

Watts, Iain P. "'We Want No Authors': William Nicholson and the Contested Role of the Scientific Journal in Britain, 1797-1813." *British Journal for the History of Science* 47 (2014): 397-419.

Wheeler, Roxann. *The Complexion of Race: Categories of Difference in Eighteenth-Century British Culture*. Philadelphia: University of Pennsylvania Press, 2000.

White, Paul. *Thomas Huxley: Making the "Man of Science."* Cambridge: Cambridge University Press, 2003.

[Winch, Nathaniel.] "On Safe-Lamps for Coal Mines; With a Description of the One Invented by Mr. Stephenson, of Killingworth Colliery." *Philosophical Magazine* 46 (1815): 458-60.

Winter, Alison. *Mesmerized: Powers of Mind in Victorian Britain*. Chicago: University of Chicago Press, 1998.

Wollstonecraft, Mary. *A Vindication of the Rights of Woman*. Edited by Carol H. Poston. New York: W. W. Norton, 1988. Originally published 1792.

Wright, A. J. "Davy Comes to America: Woodhouse, Barton, and the Nitrous Oxide Crossing." *Journal of Clinical Anesthesia* 7, no. 4 (1995): 347-55.

Yeo, Richard. "Genius, Method, and Morality: Images of Newton in Britain, 1760-1860." *Science in Context* 2 (1988): 257-84.

———. *Defining Science: William Whewell, Natural Knowledge and Public Debate in Early Victorian Britain*. Cambridge: Cambridge University Press, 1993.

Index